WHAT EVERY
HORSE
SHOULD
KNOW

WHAT EVERY
HORSE
SHOULD
KNOW

**RESPECT, PATIENCE, AND PARTNERSHIP
NO FEAR OF PEOPLE OR THINGS
NO FEAR OF RESTRICTION OR RESTRAINT**

CHERRY HILL

BEST-SELLING AUTHOR OF *HOW TO THINK LIKE A HORSE*

 Storey Publishing

*The mission of Storey Publishing is to serve our customers by
publishing practical information that encourages
personal independence in harmony with the environment.*

Edited by Deborah Burns
Art direction by Mary Winkelman Velgos
Cover design by Philip E. Pascuzzo
Book design by Patrick Barber/McGuire Barber Design
Production by Jennifer Jepson Smith

Cover photograph by © Bob Langrish
Interior photography by © Richard Klimesh and Cherry Hill,
    except for © Nathan Lake, page 123
Illustrations by © Elayne Sears

Indexed by Samantha Miller

Storey Publishing
210 MASS MoCA Way
North Adams, MA 01247
www.storey.com

Printed in the United States by Versa Press
10 9 8 7 6 5 4 3 2 1

LIBRARY OF CONGRESS CATALOGING-IN-PUBLICATION DATA

Hill, Cherry, 1947–
    What every horse should know / by Cherry Hill.
        p. cm.
    Includes index.
    ISBN 978-1-60342-713-5 (pbk. : alk. paper)
    ISBN 978-1-60342-716-6 (hardcover : alk. paper)
    1. Horses—Training. I. Title.
SF287.H554 2011
636.1'0835—dc22
                                                                    2010030073

To Zinger

Contents

What *Do* Horses Need to Know?

When watching horses, we often say, "He should *know* that . . ." — similar to what our mothers said as we were growing up ("You should know better"). Once we reach a certain age and know XYZ, you'd think we'd have learned ABC, but that's often not the case. Frequently the basics are missing, with both people and horses.

Basics are the building blocks, the foundation of everything that is to come. If there are holes in the foundation, at some point the horse, the person, or the barn could come tumbling down.

Sugar, for example, is a sweet gelding that makes a pretty picture at a horse show — *if* buddy Spice is standing alongside the rail. He also has a habit of moving during mounting, so he must be held while his rider gets aboard; he must be tranquilized before loading; he always jerks the right hind away from the farrier. . . .

Sugar has holes: things he really should know; things he should have learned at the beginning. Although he can walk, jog, and lope with the best of them, important basics are missing from his training.

Horses Are Wonderful Already

Before we even begin to make a list of what we think a horse should learn, however, let's celebrate all the splendid things he inherently knows. I've discussed in detail why horses do what they do in my previous book, *How to Think Like a Horse.* There I give you a bird's-eye view of a horse's evolution, physical traits, senses, and behaviors. Horses bring with them beauty, nobility, grace, curiosity, generosity, honesty, and forgiveness. They have amazing physical attributes, keen senses, and strong instincts, and they are very social animals. Such rich character is a great gift to us.

Wild horses know everything they need to survive. They are complete. It's when we domesticate a horse and bring him into our world that he needs to learn new things in order to adapt. As we develop a partnership with him, it is best to preserve those things that make a horse a horse. That way, there are no losers — both human and horse emerge winners. If you work together for safety, effectiveness, and unity, it will be a satisfying and successful experience.

Training Concepts and Programs

There are good step-by-step books that can help you master the nuts and bolts of horse training. This book, however, focuses on the behind-the-scene goals necessary to develop a trainer's consciousness. Understanding training concepts is helpful for seeing the big picture. You'll find that certain themes recur throughout a horse's life, from foalhood through his senior years.

Whether you are handling a foal for the first time or asking your riding horse to cross a creek, there will be elements of fear — but also, hopefully, of leadership, trust, willingness, patience, mutual respect, obedience, confidence, and harmony. Understanding this when it comes to handling, working with, and riding horses will help you become a more complete horse trainer. Understand the concepts, master the skills, develop the horse.

This book is devoted to those universal lessons that every horse should know, whether a trail horse or reiner, dressage horse or jumper, rodeo or ranch horse. Each discipline will have its own set of specific skills that he will need to learn, but all horses should know certain basics.

Throughout my life with horses, I've been both a "be here now" and "back burner" trainer. When I work with a horse, I am in the moment. Afterwards, I take a bit of time to review what happened, where we are, where I'd like to be, and what skills and principles will help us get there. Then I put all of it on the back burner until the next time I work with the horse. Things have a way of reordering themselves in the subconscious.

It is my hope that you'll read this book from cover to cover, reread parts in between training sessions, add something to the pot, and put it on your back burner to simmer. Skills are essential to master, but it's the concepts that really bring about those "ahhhh" moments. You'll see how a single training principle can be thought of separately, yet, when they all intertwine, the horse is made whole.

You can use the concepts to help you design a custom training program for each horse. The subjective and objective goals

Horses are wonderful already. Sherlock is an athlete, full of beauty and grace.

When we domesticate horses, we help them adapt and feel safe. Suckling Dickens is confident enough to leave his dam and go sloshing through the spring with me.

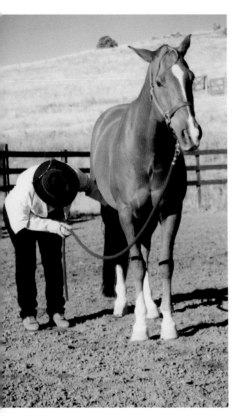

Testing a horse on the basics will show you where he needs work. Dickens knows to stand square on a long line and is patient and tolerant, but he could still use some work on his ticklish areas.

in chapter 15 will help you get started. The checklists there are designed to help you find an entry point for your horse, and they will give you some ideas of what you can review or work on next. You'll constantly be making and revising individual "to-do" lists for each of your horses.

What *You* Need to Know

Horse training is not strictly linear. At any one time, several things are occurring. In addition, each horse comes with his own set of influences: age, sex, breeding, health, soundness, condition, previous handling, temperament, and attitude. One gelding might pass one test quite easily but need more time to master another, while his full brother might respond vice versa.

As you make your plan, keep in mind your own temperament, experience, talent, timing, physical abilities, and goals.

Thankfully, horses tell us every day what they need to learn. Their gaps become quickly evident because, until they are filled, they will appear in all types of situations. A horse that has never learned to stand still, for example, might paw and move around when tied to a hitch rail, move back and forth when the farrier is trying to shoe him, sway side to side and move up and back in a trailer, or start walking while a rider is mounting. He has missed the basic lessons of *whoa* and patience.

That's why no matter what age a horse is, it's a good idea from square one to evaluate what he does and does not know. This is especially important when buying a horse. The test ride might go fine, but is he cooperative for clipping, bathing, trailering, and shoeing? By testing a horse on the basics, you'll see whether his schooling is complete or deficient.

No horse is perfect. No horse performs everything perfectly every time. Horses are living beings, not machines. Each horse comes with natural talents and challenges — some things come easily, others are tough. Our role is to fortify the strong traits in a horse's nature and help him develop so that he becomes more confident in his weak areas.

As you work toward accomplishing your goals, keep the following things in mind.

1. Break large goals into smaller, more achievable goals.
2. Do simple exercises well rather than more advanced maneuvers in poor form.
3. Be consistent. (Always be training.)
4. Be patient.
5. Preserve a horse's curiosity, willingness to learn, good attitude, and spirit.
6. Work for balance and quality of movement.
7. Let results, not time, be your measure.
8. Feed a horse according to his age and work requirements.
9. Exercise a horse daily.
10. Give a horse a job, a purpose.
11. Practice regularly.
12. Use reward and yielding to reinforce a horse's good behavior.

PART ONE
No Fear

Fear is the single most dangerous and destructive force in a relationship with a horse. Eradicate fear and you begin to develop trust.

Fear relates to one's perceived safety. When a horse sees, hears, or smells something that spells danger, he shifts into a self-protective mode. Since horses are not naturally aggressive, their instinctive response to trouble is to flee rather than to fight.

Strong fears can elicit strong responses. Horses unable to resolve their fears can panic — jump out, crash through, or rear over the threat. This can cause fatal injuries to the horse and nearby people and can damage facilities. Panic is typical of a wild or unhandled domestic horse who is suddenly confined and cornered.

Fear shows up in domestic horses that have been inadequately handled. During shoeing, floating teeth, clipping, or loading, a green horse may not try to jump over the top of you, but he may exhibit other escape-related behaviors such as striking, kicking or stomping, tail swishing, head shaking, or lunging. He is saying, "Stay out of my space" or "Let me out of here!" With such a horse, things can suddenly escalate — and when he exceeds his fear threshold, he can explode.

Systematically and sensitively assuaging a horse's inborn fears, then, is your main goal in the first phase of a horse's training. You want your horse to feel safe. You want to feel safe. You want your horse to trust you. You want to trust your horse.

Part of our job is to introduce our horses to a variety of things they might encounter in our world. Seeker now says "no big deal" when it comes to the silver umbrella, but it got her attention when I first opened it (as you'll see later in the book).

No Fear of People

★

I think an untrained horse fears man's world
more than he fears man himself.

★

FEAR IS A REALITY WE have to deal with when handling horses. When fearful, horses can be anxious and uncomfortable and can become unpredictable and dangerous. That's why we must help them deal with their inherent fears.

The first fear we must address is the fear of humans. Fortunately, a horse is born with everything he needs to overcome his fear of humans: curiosity, a strong power of association, deeply rooted social-hierarchy behaviors, a need for security, and a wonderfully generous nature.

Putting a hand on a horse is the most basic means of connection and communication. Sassy has learned that being touched is no cause for concern and is actually a time to relax and luxuriate.

If you can show a horse that he can rely on your good judgment and leadership, this will be the basis for a great partnership.

Predator–Prey?

A common explanation of why horses are naturally fearful of humans is that humans are predators and horses are prey animals. Although a modern domestic horse has inherited some of his wild ancestors' fear of being hunted by animals for food, to me it seems unlikely that this would be the main reason horses fear humans today.

I think an untrained horse fears man's world more than he fears man himself. Unless a human behaves like a predator — that is, chases, corners, and threatens — a horse will most likely be more curious than afraid. When a wild or untrained horse does see a person, he will understandably be alert and most likely suspicious of the unusual two-legged animal, just as he would be of any other unusual creature, such as a bear. But put that same unhandled horse in a small enclosure, add some tack and other training paraphernalia, and apply a little bit of human pressure: you now have a recipe for fear.

Touch and Trust

To reduce a horse's natural fears, we must first convince him that we won't hurt him. This is the basis for trust. Next, we must show that we are a wise leader. We must be fair and not ask him to do something difficult before he is ready. Never ask a horse to do something that is unsafe. One primary fear a horse must overcome is that of the human touch. I mean, simply, hand on horse, before any halters or tack come into the picture. Not only must a horse allow you to touch him in a safe zone, such as on his neck, but eventually anywhere and anytime, from the front, the back, and both sides. Then, not only must he allow you to touch him all over, but he should also be relaxed about it and hopefully learn to enjoy your touch. For your safety and his, there should be no sudden *eek!* moments.

Safe Zones and Hot Spots

Horses are very tactile creatures that naturally enjoy and solicit being touched in certain places, such as the thickly muscled areas of the neck, chest, and rump, and the withers and back. They do not naturally enjoy touch in the thin-skinned

When late-term Sassy was turned out on pasture one early spring day to mosey and stretch for an hour, a cold, drenching Rocky Mountain hailstorm suddenly came down. I ran out to bring her in, but she had already begun the process of foaling. Richard and I assisted her out in the pasture, and by the time the hail stopped, the foal was out of the sack. I tied up the placenta and led Sassy while Richard carried the vigorous foal up the hill to the barn. All four of us were soaking wet, so we stood under a radiant heater to dry while we toweled off the mare and foal.

The foal, less than an hour old, walked all over the barn, exploring every stall and nook and looking for clues in his new world, so we immediately named him Sherlock. All the while, he carried an oversized bath towel across his back, as if it was routine stuff. He arrived in the world of humans without fear, a hale and unique foal.

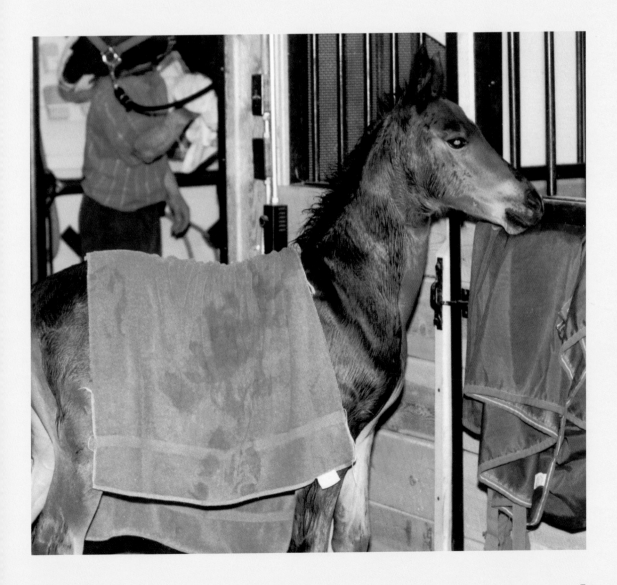

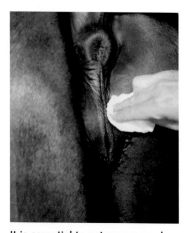

Beginning on foaling day, handle a foal's mouth to prepare him for veterinary care, deworming, bridling, and more.

Lift and Don't Lift

There are times when you want your horse to lift his feet, such as when you want to clean them or when your farrier wants to trim or shoe. But there are also times when you want to handle your horse's legs and not have him move his feet. When you bathe, groom, clip, bandage, or put on leg boots, you don't want your horse to lift his feet, as it would be not only counterproductive to what you are trying to accomplish but dangerous as well. Therefore, every horse needs to know when to respond to your request to lift his feet and when to stand patiently with his feet firmly planted on the ground while you perform procedures.

It is essential to get a mare used to having her vulva washed and touched for breeding preparation and procedures.

Prior to saddling and riding, accustom your horse to being touched all over — they love solid body-to-body contact.

areas such as the flank, sheath, udder, anus, mouth, and ears. Thus it is our job to systematically help a horse overcome his aversion to being touched in those "hot spots." You'll eventually need to wash the udder and sheath, take his temperature, and not worry when the farrier's hat tickles your horse in the flank. Through a system of progressive desensitization, a horse will realize that he will not be harmed no matter where he is touched.

For riding, you'll need to be able to put your leg on your horse without him jumping out of his skin. (See box, pages 8 and 9.) Many horses are overly sensitive to leg cues because they were never properly desensitized to touch and pressure on their ribs. Because you want your horse to retain responsiveness to your leg aids, you don't want to overdo the "sacking out" of the horse's barrel, but a horse should feel comfortable with you touching and rubbing his sides.

Horses prefer a firm assured touch rather than a tenuous tickle. I've known quite a few horses who are much more content with a rider astride than with the same person on the ground. A valuable lesson for every horse is for the rider to mount the horse bareback and rub, hug, lean forward, lie back, and otherwise maximize body-to-body contact with him. It is a grounding exercise for both horse and human. (See opposite for a diagram of touch points.)

TOUCH POINTS

A horse's body has various zones of sensitivity. Some areas have thin skin, very little protective muscle tissue, and nerves and vessels close to the surface. Those areas are generally more sensitive than regions with heavy muscles and thick skin.

Sensitivity varies among breeds and individuals within a breed, but this Touch Points map will give you a general sensitivity guideline. It will help you choose the proper touch and grooming tool and give you the basis for interpreting your horse's reactions and planning your aids and cues.

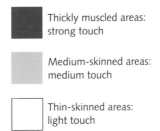

Thickly muscled areas: strong touch

Medium-skinned areas: medium touch

Thin-skinned areas: light touch

Mounted exercises are designed to give a rider confidence on a horse's back and to provide an added measure of security if one must dismount in a hurry. But the horse also needs to be familiarized with the sensations and activities of a rider. Unlike the experienced, "bombproof" school horses we learned on, horses are not naturally tolerant of strange sensations and antics in their blind spots. A green horse might be confident as long as a rider is properly astride; if things get goofy, though, and someone is hanging half off one side, with the saddle slipping, some horses will go into a real snorting, bucking, fearful fit.

As part of every horse's education, therefore, these two exercises (and more like them) should be practiced regularly so that the horse is comfortable and accepting of odd positions and movements all around him, especially in his blind spots.

FLYING DISMOUNT

One of the first exercises taught to me as a young rider was an emergency dismount, also called the Flying Dismount. Various versions depend on whether you are riding bareback or with an English or Western saddle, but it basically involves taking your feet out of the stirrups, leaning your upper body forward, and pushing up and off as you swing one leg over the rump to land on both feet on the desired side. We were first taught to master a flying dismount at a standstill off both sides, then at a walk, trot, and canter, off both sides.

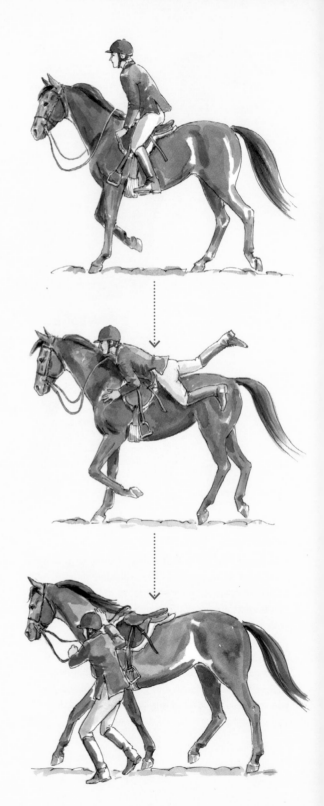

AROUND THE WORLD

During the same era, I was taught Around the World. The unflappable school horse was happy to stand still while I swiveled in the saddle from a normal astride position to one with both legs on the near side to a rear-facing position to one with both legs on the off side to a return to the normal forward-facing position. I had gone "round the world."

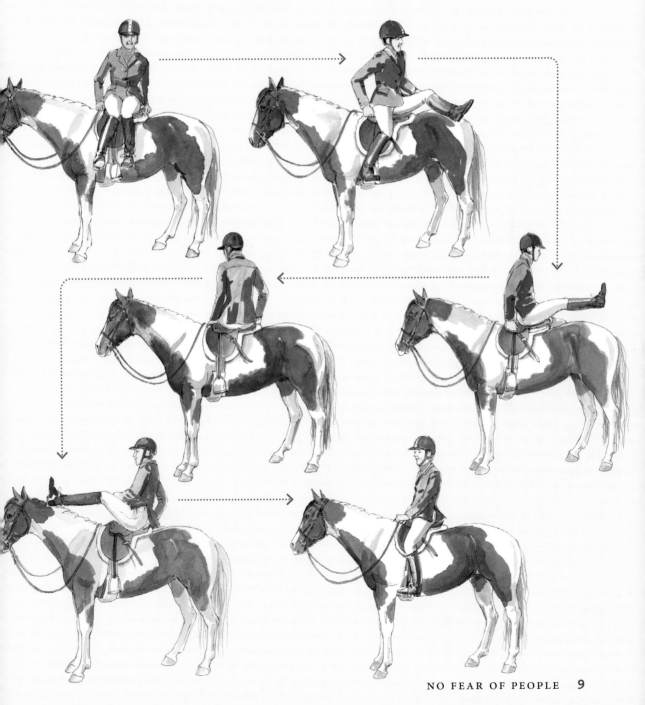

Fear of Someone in Blind Spot

Horses are acutely aware of their surroundings, and because of the location of their eyes and their wide range of vision, they are able to pick up the subtle movements of things all around them. They do have blind spots, however — areas where they can't see what's going on — and that can make them nervous. It's similar to when you are driving your car or truck: sometimes you can't quite see the vehicle (or even if there *is* a vehicle) in that spot between your rearview mirror and your sideview mirror. In some situations, you might be tempted to turn quickly and look at that spot or slow down so that if there is a vehicle, it will pass you.

In a similar way, a horse is uneasy if he senses something to his side or rear that he can't see. He is more comfortable if he can see a thing clearly and head on. If there is a perceived threat off to the side or the rear of a horse, he will most likely try to turn and face it.

You've been taught from day one that when you are behind a horse or approaching him from the rear, you should let him know of your presence by putting your hand on his hindquarters. Well, this needs to be expanded to allow you to be able to touch your horse when he is in any position, any time, anywhere. As always, these suggestions need to be implemented with common sense for safety's sake.

A horse needs to be comfortable with you:

- Next to his head on the near side.
- Next to his head on the off side.
- At his girth on the near side.
- At his girth on the off side.
- In front of his head.
- At his hindquarters on the near side.
- At his hindquarters on the off side.
- Behind him.
- Standing above him in all positions.
- Squatting below him in all positions.
- Farther away, doing all of the above.
- Mounting, and moving around on his back while mounted.

When approaching a horse from the rear, realize that when you are directly behind him, he can't see you. Zipper's prominent eye allows him to see me taking his photo as I approach seven-eighths rear. His ears are tuned back to me and he is experienced enough to know there is nothing to fear and yet, out of reflex, he still takes one step forward with his right hind.

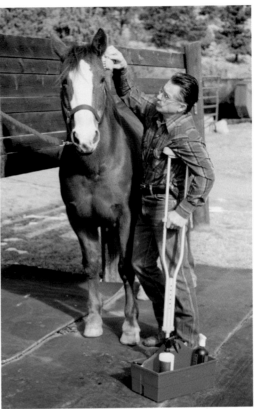

Eventually, a horse needs to be comfortable with all sizes and shapes of humans: very tall or short; toddlers and infants in arms; elderly or injured people walking oddly or with crutches, walkers, or wheelchairs; kids running and leaping.

Catching

Your training relationship with your horse really takes on a formal tone when you go to catch him, so it is a perfect time to start things out on the right foot. In an ideal situation, a horse should know that when you stand at his gate with a halter in hand, you want him. He should respond by coming to you, with or without a signal or cue. But it would also be satisfactory if he just turned and faced you when you entered his stall, pen, or pasture, and stood still while you approached him. It would be unacceptable if a horse turned or moved away from you when you went to catch him.

It is a good sign that when a horse sees you coming to the gate, he starts walking toward you. Seeker triggers off the fact I have a halter in hand and have headed to her gate, so she walks up to meet me and be haltered.

How (and How Not) to Catch a Horse

Here is where you might be thinking, "My horse avoids being caught. How do I change that?" The best way to develop and maintain a horse's good attitude about coming when called, or at least facing and standing when you approach, is to have a pleasant attitude yourself and to do something pleasurable to him when you first catch him. It can be a simple rub on the forehead, brushing flies off his chest, giving him a scratch where you know he likes it, making a certain sound, or even using a treat the proper way. All of these reward a horse for being caught and make him look forward to being caught in the future.

If you want to teach a horse to fear or dislike being caught, give him a slap on the neck when he finally does let you approach, halter him roughly, give a jerk on the lead rope when you finally get him haltered, and generally be in an ill temper yourself. That should do the trick!

As with so many other interactions with your horse, during the catching and haltering procedure, he is learning, so take your time as you catch him and develop a good association. When I'm with a horse who has learned to dislike catching and haltering, I might carry a halter and lead rope with me, approach him, give him a rub, but not halter him; instead, I just walk away. Or I might catch, halter, rub, remove halter, turn loose, and walk away. Or I might just hang out with the horse for a bit with no agenda.

How a horse is turned loose is also very important and will be covered in chapter 5.

No Fear of Humans

Test note: The tests at the end of the chapters are meant to exemplify the concept discussed. The individual tests are by no means a complete checklist of the concepts discussed, but are provided only to give you an example of what your horse should know. For an extensive list of exercises, refer to chapter 15.

Test A. Enter your horse's living quarters. Your horse should walk up to you or turn and face you and stand still while you approach.

LEFT: Enter your horse's living quarters and stop. Ideally your horse will walk up to you as Seeker is doing here. This comes in especially handy in a large pasture.

BELOW: Alternatively, your horse might stand and look at you while you approach. That is a fine option but might require a lot of walking if you are out on pasture. For demonstration purposes, in these two photos I stop Seeker square and ask her to stand there while I approach three-quarters front. You can tell it is all she can do to keep from coming to me, as that is her normal MO.

Test B. Approach your horse. Touch him on the neck and then run your hand along his side and under his belly all the way to the udder or sheath. Walk around in front of him to the off side and rub his right ear and then walk back and lift his tail so you can see his anus. If he's wearing a sheet, adjust the rear leg straps. Your horse should be as relaxed and unconcerned with your position and handling as Seeker is.

No Fear of Restriction or Restraint

Restraint shouldn't break a horse, it should build his confidence.

ALTHOUGH RESTRICTION AND RESTRAINT ARE related, there are some important distinctions between the two when used in horse training and management. The term **restriction** refers to limitation. Although it usually means limitation of movement, it can indicate other types of restrictions as well. **Restraint** means stopping altogether — usually preventing some type of movement.

Every horse must learn to accept a certain amount of both restriction and restraint to live and function safely in the world of humans. The manner in which the restriction and restraint are introduced and applied, however, will determine, to a large part, how the horse reacts: with acceptance, or with fear.

Restriction

The classic example of restriction of the domesticated horse is confinement. When we put a horse in a pasture, paddock, pen, stall, stocks, or trailer, we are restricting his movement, his socialization, his eating, and more. Each successively smaller enclosure restricts him further. If a wild or untrained horse is put into a stall "cold turkey," he can react quite violently and dangerously. Starting him out in a large and safely fenced pasture and gradually decreasing the size of his enclosure, however, will accustom him to the restriction gradually.

Socialization and Fear of Separation

Two types of restriction related to confinement are the isolated life and the forced socialization of domestic horses. A wild horse, subject to sexual and herd behaviors, is "free" to choose his companions. I say free loosely because anyone who has observed wild horses knows there is a lot of fighting involving reproductive hierarchy that includes or excludes certain horses from certain bands. With that said, wild horses are somewhat free to choose their associates and rarely live alone.

The domestic horse, on the other hand, is usually not free to choose where he lives and with which other horses. He might be required to live alone or to share quarters and feed with

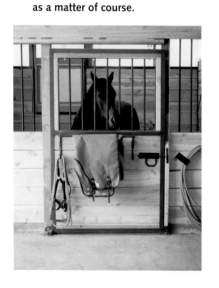

A horse that is not afraid of being alone lives in a stall calmly and contentedly. Aria, who normally is housed in an individual pen or in a herd on pasture, was born in this very stall and accepts confinement as a matter of course.

groups of horses that may or may not be compatible. Thankfully, horses most often are quite adaptable and can learn to live alone. And they can usually work out acceptable routines with one another once a pecking order has been established (see chapter 5).

For a horse's safety and ours, every horse needs to learn to accept isolation from, as well as social interaction with, other horses. You will need this acceptance to be able to care for and use your horse. It affects everything from veterinary and farrier care to trailering to riding a horse alone or with a group.

The thoroughly socialized domestic horse is content living solo (left) or with other horses (right). This bay horse is content in either situation.

Separation Anxiety

When a horse is anxious, nervous, afraid, or just not accustomed to living away from other horses, he can become **herd-bound, buddy-bound,** or **barn-sour.** These are related but different behaviors that can be grouped together and thought of as **separation anxiety.**

This insecurity causes a horse to misbehave in various ways when separated from a group of horses (the herd), from a single buddy or preferred associate, or from a physical location that represents security (the barn or feeding area). The misbehaviors are attempts to return to the herd, individual, or barn that represents comfort and safety.

Tight, dark spaces can seem dangerously restricting to horses. That's why entering a trailer for the first time can elicit fear. This rock crevice is only wide enough for Zinger and me to get through with an inch or two to spare on either side. After years of trust, she enters without hesitation.

Domestication places restrictions on horses in many ways. From space confinement to social isolation to tack that limits movement, a horse must learn to accept restriction.

Confinement posed by living quarters
★ Pasture fence
★ Paddock
★ Pen
★ Stall

Other types of space confinement
★ Trailer
★ Stocks
★ Cross ties and hitch rail
★ Narrow trails, spaces, low ceilings, low branches
★ Entering narrow spaces, dark spaces, low ceilings

Social restriction
★ Separation from other horses
★ Forced socialization with other horses

Restriction by tack
★ Halter
★ Bridle
★ Girth or cinch
★ Rear cinch
★ Breast collar
★ Crupper (for driving, packing, or riding in the mountains)
★ Blanket, fly sheets, fly masks, hoods, boots

Handling and riding
★ Regulation of speed
★ Regulation of direction

Separation anxiety is easier to prevent than it is to cure. Its main cause is a fear of being alone — insecurity. A horse who does not feel safe or content alone or in the world of humans is more apt to act out in a desperate need to be near other horses, often a preferred associate or buddy. Preferred associates are those you might see standing head to tail with each other and performing the satisfying ritual of mutual grooming.

Safety in Numbers

It is natural for horses to want to be near each other because they are gregarious animals that, over the last 60 million years, have learned that they are safer and more content in a group. When you are a prey animal, there is safety in numbers. Also, the strength of the bonds among various herd members contributes to herd safety and, in part, allows the herd pecking order to function. Foals initially have the strongest ties with their dams. As foals grow older, they look to other herd members to satisfy their need for social and sexual interaction.

Wild horses and horses who live in groups participate in many social customs such as mutual grooming, grazing, and drinking; playing for physical and sexual development; and sharing guarding and sleeping routines. It is important to allow

TOP: Horses are highly socialized herd animals and prefer to live in groups for safety and confidence.

ABOVE: Although it is somewhat contrary to an equine's herd nature to live alone, a horse can adapt very well to being the only one on the farm as long as he has social interaction with humans or other animals.

horses to "be horses" — to see, hear, smell and, if possible, interact by touch with other horses. Many show and performance horses are not afforded these luxuries because life in a herd can produce negatives such as chewed manes and tails, bare spots from mutual grooming, and wounds from biting and kicking. And because land is dear, we often have to keep our horses on a small acreage or in boarding facilities where horses are often, out of necessity, housed in individual stalls or small pens.

For better or worse, therefore, domesticated horses, in order to be comfortable and useful in the modern world of humans, must be made to feel secure eating, sleeping, and being handled and ridden, both among various groups of horses and without other horses in close proximity. Early handling by humans makes a life-long impression on a horse. According to behavioral studies, foals that are handled early are more curious and investigative than unhandled foals. When handled foals are startled or alarmed, their fear disappears more quickly than that of unhandled foals. It is important that early handling is founded on sound principles of horse behavior and training. Poor handling or overhandling is worse than no handling at all.

Foals that are handled early are more apt to leave their mothers' sides, play farther away from their dams, and show more self-confidence. In contrast, unhandled foals are generally reluctant to leave their mothers. Very early on, Sherlock was ready to leave his dam's side and explore.

Prevent Insecurity by Early Foal Handling

Foals are greatly influenced by their dams' personalities. A foal with a very timid mother may inherit and develop a timid manner and stick close to her side, remaining very dependent on her. Foals with self-confident dams, on the other hand, are often curious and experimental and interact with all sorts of other creatures with great interest. Such foals usually have a strong independence aptitude early in life. The combination of healthy self-confidence and curiosity results in an ideal horse to train.

FIRST DAYS AND MONTHS

Early handling begins with bodily restraint of the day-old foal. This is an essential lesson to teach him that there is no danger in being contained and momentarily separated from his dam. Gradually increase the periods of separation during the first four months and by weaning time, the final separation will go smoothly.

A foal should be touched all over so that he learns to accept and not fear humans. "Grooming" a young foal by scratching him on the withers and over the tail head, areas where a foal particularly likes to be rubbed, further convinces him that the human touch brings enjoyment and contentment.

ONWARD TO IN HAND

In-hand lessons build up a horse's confidence and experiences by helping him conquer fears as he is led over, under, around, and through safe "obstacles" of all sorts. This will encourage his curiosity and independence and cement a solid working relationship with his handler.

How Horses Act Out

Some separation anxiety behaviors include (but are not limited to): balking (refusing to move forward), refusing to leave a specific area, running toward home, racing to catch up with a horse ahead, wheeling around suddenly, swerving, rearing, spooking, carrying his body crookedly, trying to back up or turn around, becoming gate-sour in an arena, moving short and quickly, getting behind the bit, screaming, pawing, stall walking or pen walking (pacing back and forth), and pushing at or through fences or gates. Some horses that can't cope with their insecurity when confined can form vices such as cribbing or stall kicking when separated from other horses.

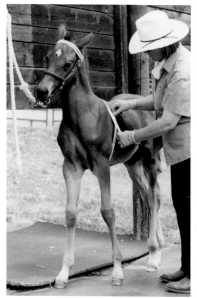

LEFT: If the first handling lessons are fair and well thought out, a foal will have a basis for trust and confidence in humans and will look forward to interacting with people. If, on the other hand, the first handling is inappropriate and rough, the foal will have no incentive to develop a relationship with humans. Suckling Sherlock accepts tying and is not worried about the touch and the girth restriction of the weight tape.

RIGHT: Early lessons pay off. Handling weaning Sherlock in the pasture is just as easy — he's just bigger!

WEANING: A MAJOR MILESTONE

Weaning is a marker event and a major step toward independence. Weaning can be very anticlimactic or traumatic. (See more about weaning in the box on page 34.) If a suckling foal has been handled properly and in a variety of situations, he will be mentally and emotionally ready to face weaning time. Separation from the dam will occur as if it is just another day in a progression of days.

If a vulnerable weanling is managed improperly, however, he might shift into a stressful state of panic. If a foal is not handled before weaning and then is weaned abruptly, his anxiety may turn into fear or result in injury. If he is weaned and put with other foals or horses and not handled, he may form such a strong dependency on one or more of the other horses that he has to go through a type of weaning all over again when separated from them.

That's why it is important that, throughout the young horse's life, he receive interaction with his human handler in the form of training. Regular handling will develop confidence and independence in the foal and will be the basis for a human-horse partnership.

Dealing with Separation Anxiety

When a horse fears restriction and separation be sure to evaluate all management areas and adjust feed rations and exercise routines if necessary.

BUDDY-BOUND DOS AND DON'TS

If a riding horse is buddy-bound, you might feel that it's the safest thing to always ride next to the horse's best pal. But actually you are reinforcing the horse's insecurity and creating an accident looking for a time and place to happen. At

TOP: Giving a young horse interesting things to do allows him to develop confidence. Here ten-month-old Blue calmly walks like a veteran over irrigation plastic.

BOTTOM LEFT TO RIGHT: One of the first separations of a foal from his dam might occur with a creep feeder. Sherlock has no problem leaving Sassy to enter a narrow passage and duck under a low bar to reach his ration.

Reward Calmness, Not Anxiety

Without knowing it, you might reinforce undesirable habits in a horse by rewarding him when he acts out with negative separation behaviors. For example, let's say you turn one horse out to graze and leave the other in his pen. If the horse left behind begins bucking, kicking, pawing, and screaming because his buddy has disappeared over the crest of the hill, it is not uncommon for a person to go up to him and talk baby talk; pet him; feed him; or, worst of all, cave in and turn him out with his buddy.

All of these pleasantries have, in effect, told the horse: "Screaming and pawing are productive behaviors. They bring rewards of feed and freedom." The next time that horse is separated from his buddy, he will once again paw and scream because, why not? Last time he got the goods. In fact, he is likely to paw and scream *worse* than he did the first time until he gets what he wants.

The best thing to do is nothing. As long as the penned horse is in a safe enclosure, leave him. Eventually most horses learn that they can survive safely alone, and the acting out will disappear.

---

some point, the two horses will have to be separated. Either the buddy horse will move ahead or lag behind, or one horse might have to stay behind while the other is ridden to get help in an emergency. A horse should be able to be ridden safely alone or with a variety of other horses.

Herd-bound horses are emotionally tied to the herd in general or to individuals within the group and will react much like a buddy-bound horse when being separated from the herd.

BARN-SOUR PREVENTION AND CURE

The combination of healthy self-confidence and curiosity results in an ideal horse to train.

Whether or not a horse is barn-sour, instead of starting a ride at home, trailer to a park or trailhead or to another arena or stable for your ride. The lack of the familiar starting point may decrease old habits and will most likely make your horse listen and look to you for guidance. This is your chance to be assertive yet fair and to build your horse's confidence in doing something new.

When you head out, trot or lope energetically. On the way home, though, always walk him, especially the last mile. Lateral work, such as leg yielding or two tracking, is also good on the way home: it relaxes a horse and slows him down, because he must step deeply sideways and underneath his belly as he moves forward.

Many barn-sour horses quicken the pace the closer they get to home. Such a horse should be stopped at various points along the way. Stand for a few minutes, either facing or turned away from home. When you reach your driveway, walk past it several times before you finally turn in.

Do the same as you approach the barn. Don't make a beeline for the barn and then hop off, loosen the cinch, and give your horse an apple. That will just teach him that the barn is the best place in the world and to hurry home even faster next time! Instead, ride back and forth past the barn a couple of times, take your horse back out on the trail for ½ mile (0.81 km), or head out to the arena and lope quietly for about 5 minutes. Then dismount and lead him to a strong, safe tying place away from the barn.

Let him stand there saddled (with the cinch loosened a few holes) for as long as it takes for him to relax and lower his head. Once your horse is standing quietly, untack, groom, and care for him and return him to his stall or pen.

LEFT: Develop a partnership with your horse and all sorts of trails will open up for you. Zinger and I went on many solo adventures together.

RIGHT: It is equally as important for a horse to be comfortable being ridden with other horses but not be glued together. These endurance riders can ride safely next to or out of sight of one another.

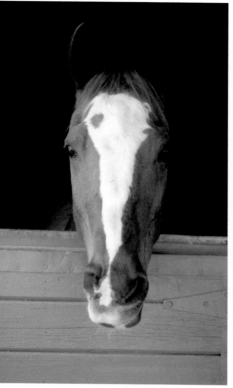

LEFT: When a horse is cooped up, it is a perfect time to bond with him to give him the socialization he desires from other horses.

RIGHT: Training includes restriction. Long yearling Blue is restricted by tying and feels restriction from cinch pressure, yet she stands calmly at the hitch rail.

Using Management to Increase Security

Some aspects of management can affect a horse's level of security. A cold, wet, sick, or very hungry horse is more apt to be physically anxious so will most likely be emotionally unstable. Sometimes, just making a horse comfortable will make him content enough not to worry about his confinement or the whereabouts of his companion. On the other side of the coin, an overfed and underexercised horse can develop frisky or pushy habits similar to those of a herd-bound horse even if he isn't emotionally insecure. That's why individually managing each horse's feed and exercise is essential.

Training Includes Restriction

Ground training, driving, and riding contain all sorts of restrictions. We limit where a horse goes, how fast he moves forward, at what gait, at what speed within a gait, the form in which he carries his body, and whether he moves forward, backward, sideways, or in a combination. All of ground training and mounted training is a type of restriction — we are shaping a horse using various methods and techniques.

Reward or Bribe?

There is a big difference between setting up a training session during feeding time and feeding a screaming horse to placate him. Say you plan to work each day for 2 weeks on minimizing your horse's anxiety at being separated from his buddy. At each feeding, feed the insecure horse and then take his buddy as far away as you can until the insecure horse gets anxious. Stop at that point and wait until the insecure horse relaxes and resumes eating. Then, return the other horse to his pen or stall. The next feeding, repeat the procedure seeing whether you can walk a little farther away. Using feeding time this way to approach the problem is entirely different from using feed to bribe a horse.

Now, here's where feed is used as a bribe. It is the middle of the day and you take a horse's buddy away. The horse left behind begins fretting and squalling. You want to shut him up so you dump a scoop of grain in his feeder to try to placate him. You have rewarded him for his bad behavior.

The difference between the two uses of feed is similar to the difference between training a horse to load into a horse trailer and bribing him. If you tempt him to walk in by leading him forward with a bucket of grain in front of his nose, you've bribed. If you teach him to walk in on command and let him find that there is feed waiting for him in the trailer's manger, or if you put feed in the manger once he is loaded and locked in, you've trained him using positive reinforcement.

Adapting to Tack Restrictions

A surcingle or saddle girth or cinch restricts a horse's breathing somewhat and, in that way, the girth pressure is a restriction. When a cinch is too tight it can be uncomfortable, but even when it is tightened just snugly, the horse receives a restrictive sensation — much like when your belt is too tight after a big meal. Cinching up fast and rough will elicit an *oof* reaction from the most seasoned of horses, let alone a youngster or untrained one. So first fasten a cinch safely snug, and then progressively tighten it until it is suitable for mounting and riding. This rarely results in a bad reaction and is considerate of the horse.

If your facilities and schedule will allow, think of ways you can let your horse "be a horse" from time to time. By this I mean allowing him the opportunity to exercise, rub, roll, mosey, graze, and socialize with other horses.

Even though turning all horses out together on a large pasture is often thought to be ideal, it may not be possible, practical, or even desirable to do so. There is the normal scuffling of herd hierarchy as well as the inconvenience and safety considerations when you must go out on a large pasture to catch one horse amid a large group. And once they have lived in individual pens, some horses may enjoy a brief turnout with other horses but seem to prefer the peace and quiet of an individual sheltered pen, where they can eat alone and have respite from sun and bugs.

Get to know each of your horses and provide them with an appropriate amount of turnout with or without grazing, a place to roll, and some time with compatible horses.

Turnout provides the opportunity for moseying, grazing, rolling, and socializing. These Wyoming horses are in an ideal setting.

Other tack items that cause restriction include a rear cinch, noseband, crupper, tie-down, martingales, breast collar, and bridle.

Limiting Sensory Stimulation

Restriction of vision is an element that can cause a horse distress. When he is put in a stall, trailer, or other enclosure, he

has limited visual access to other horses, people, and his environment. This can be unsettling until he becomes accustomed to it. Grills and Dutch doors are used in stalls, and screens and drop-down doors in trailers, to allow air and light to come in and a horse to see out.

When a horse is being ridden and the level of his head and neck is restricted, such as by a tie-down or martingale, he often can't see where he is about to put his feet. This can be unnerving when he is negotiating an obstacle, crossing water, or going over a jump. By avoiding tack that restricts a horse's ability to raise or lower his head as needed to focus, and through progressive training, you can help him overcome this fear of vision restriction.

On the other hand, it is sometimes more soothing for a horse *not* to see something. For example, if you must evacuate an area where there are small patchy grass fires, it might be easier to lead a blindfolded horse than one who can see the flames and spooks at them.

Similarly, if loud noises rattle your horse at a show, you can put cotton in his ears to restrict his hearing. This often calms a horse, but it will also make him less able to hear you and less able to process the sounds he needs to pay attention to.

TOP: By not restricting a horse's vision, you will allow him to allay his fears. When approaching a wooden bridge, I allow Seeker to lower her head and get a good look and smell as she walks on.

BOTTOM: When you put a grazing muzzle on your horse and turn him out on pasture, you have greatly reduced his ability to graze, so you have restricted but not restrained him from grazing.

Restraint

Restraint means stopping (not simply limiting) a behavior. In horse training, that usually means we stop a horse from moving, but there are other examples of restraint that relate more to management.

In very simple terms, we put a restraint on a horse's eating when we keep him off pasture and house him in a pen or stall. We might do this so he won't get too fat or founder from rich grass. Left on his own, a horse would graze continuously with only short rest periods. That's why, often for the horse's own good, we stop him from grazing.

His desire to graze is still there, however, so we need to use strong, safe building materials and construction methods to enclose him in his paddock, pen, or stall, and protect his health and safety. And we need to provide him with water, salt, and grass hay in lieu of rich pasture grasses. I intentionally started with this as a positive, care-giving example because restraint often gets an undeserved bad rap.

The Benefits of Restraint in the Real World

A horse needs to accept various types of restraint. Without the benefit of leg-restraint lessons, for example, a horse caught in wire or tangled by other means can panic and quickly injure himself severely, if not fatally. All of my horses receive thorough leg-restraint lessons so that they do not panic when they get a foot caught. This restraint training also makes it much easier for the farrier and veterinarian to handle the horses' legs. In several instances, this training really saved the day.

Restraint is a godsend during unusual circumstances, such as when a horse rolls and gets caught under fence panels or is trapped in a trailer rollover. Past restraint lessons seem to have a calming effect on a horse in trouble.

The more we can help our horses overcome their fear of restraint, the less likely it is that they will overreact during a panic-prone situation. We are, in essence, protecting a horse from his own dangerous instincts and reflexes. And don't forget that a 1,200-pound (550 kg) horse in a panic can be extremely dangerous for all humans around as well, so restraint is necessary for your own safety too.

Examples of Restraint

★ Handler uses halter to control and direct movement

★ Horse stands on three legs with fourth leg restrained by hand, rope, or equipment

★ Horse stands tied quietly at a hitch rail, trailer, in cross ties

★ Handler picks up each foot

Practice putting your hands on your horse anytime, anywhere. Approaching Sherlock in the pasture and without haltering, I work on his least favorite touch point — his mouth.

Restraint can be good, bad, or downright ugly. Improperly applied, restraint can be very dangerous to both horse and handler — that type is inappropriate. There are many instances, however, when restraint is essential to a horse's well being, so understanding the various types of restraint and the application rules will help you work with your horse in a positive manner.

We all want a cooperative horse. When a horse is in a relaxed, trusting state, he is much easier to handle. But even the most experienced horse, when faced with trauma or injury, is likely to panic if he has not had formal restraint lessons ahead of time. Restraint lessons are designed to teach a horse that it is okay for us to take away his ability to move by restraining him, his legs, or his head. He needs to know this for the veterinarian and farrier to do their work. It is far easier and safer to stitch or shoe a horse that stands still.

Methods of Restraint

Horses are subdued using various means: manual, mechanical, psychological, and chemical. Often, more than one method is used at a time.

MANUAL

Manual restraint utilizes your body position and the application of your hands on the horse. You can use manual restraint in a strategic, overpowering way or in a soothing, persuasive manner.

A horse can often be calmed with hand rubbing. Areas that respond to this technique are the forehead, the spot just below the eyes, and the area just in front of the withers. Each horse has

Your veterinarian will appreciate a well-mannered foal when it comes time to work on him. Dr. Stacey Tarr gets ready to lay Sherlock down for gelding.

Zinger Keeps Her Cool

Prepare a horse ahead of time for situations that could lead to panic. I hobble and ground-tie Sassy (left) before walking around the arena picking up rocks. Later I remove the hobbles and again require her to stand (right). These lessons pay great dividends when you need them.

During the first winter we lived in northern Illinois, I had to turn my horses out for exercise in a bare, 75-acre corn field with square wire fencing — admittedly not the best for horses, but at the time better than no turnout. Most often the three mares would graze along the edges of the field, sorting through the dead grasses.

One day, while warming up with a cup of coffee and watching them from the kitchen window, I noticed that Zinger seemed to be staying in one spot while Sassy and Poco had moved on. That seemed odd, so I got out the binoculars and saw that Zinger's left front shoe was caught in some wire. In seconds, I was out the door and across the road, then slowed down and talked to Zinger as I calmly approached her.

To my relief, she had not struggled: once her shoe had become snagged, she just stood there. For me, it was a simple matter of slipping the wire out from between the heel of the shoe and her hoof. Had she tried to free herself, the most likely result would have been that either the shoe (or possibly part of her hoof) was ripped off or that a section of fence was torn off the posts and then tangled and dragged behind her.

When I set her hoof down, she looked to me for permission to go. I gave her the sending signal, said "OK, walk on," and she was off to join her herd mates.

Same horse, fast forward 10 years. I was exploring the Colorado foothills northwest of our home with a friend. We were crossing an abandoned homestead near the Roosevelt National Forest. Because there were no trails, we made our way through the places of least resistance, and even at that, we were often up to our horses' bellies in brush.

All of sudden Zinger just stopped. Very out of character for her, and something she'd do only if she desperately had to urinate; yet she didn't raise her tail, she just stood there. I looked underneath us, and there were the remains of an ancient, rusty, barbed-wire fence roll.

When I dismounted, I had to step right into the mass of wire, which caused it to move somewhat around Zinger's front legs, yet she stood there while I removed the wire cutters from my saddle bag and cut her free. It wasn't just a snip or two. I worked for about 5 minutes, rolled the wire up and off to the side, and then asked her to back out of it.

After a thorough inspection of her legs, I was relieved to find no wounds. Her calm disposition and training got the credit. I remounted and we continued on our adventure.

No Hoof, No Movement

When you take away a horse's ability to move his feet, you have restrained him. This is the most threatening form of restraint from a horse's viewpoint.

The simplest example of leg restraint is lifting up a hoof. It's best to start this lesson with a young foal. If a horse is older and larger and hoof handling has not been covered, it can be more difficult. In any case, every horse will eventually need to learn to surrender each of his feet to a handler and stand patiently. You can teach the lesson at first by using your hands or a rope to lift up the foot (see chapter 3 for more on rope handling).

In addition to letting you lift up his foot and hold it, a horse must allow you to place the foot down. It is a reflex action, when he feels his foot being lowered, to snatch it away and put it down. You will need to override this natural tendency and teach him to let you place his foot on the ground in a slow, measured manner. Never just drop a foot when you are finished with it, as that could lead to the bad habit of the horse jerking his foot away, stomping, or kicking.

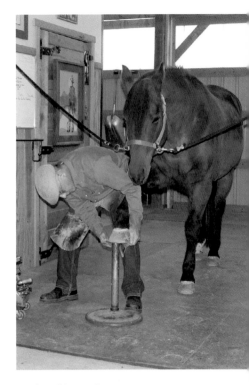

A relaxed horse that accepts restraint and restriction is so much easier and safer to handle. Seeker is very relaxed when Richard attends to her hoof care because she has been prepared with restraint and restriction training.

his own "yes" and "no" areas, however, so know the "yes" areas ahead of time so you can use them when needed. Touch one horse's ears and his eyes go wild and his head goes skyward; touch another's ears and she lowers her head and closes her eyes.

There is also a big difference between grabbing an ear and rubbing an ear. When necessary to quickly dominate a horse, a popular historical choice was to grab an ear. Although this can be effective if properly applied, it is tricky to do right and can have some very undesirable side effects, such as the horse becoming headshy, lop-eared, or even more uncontrollable because of fear.

A better hands-on alternative in an emergency is to pinch the tip of the ear between the thumbnail and another fingernail. Another option is the application of a shoulder twitch, which is grabbing a handful of skin at the shoulder and rolling the skin forward and underneath the knuckles.

Weaning

Weaning is restraint. When we stop contact between a mare and foal, we are restraining both of them. Usually, we use the restriction of a pen or stall to accomplish this restraint, but basically, we are preventing the mare and foal from contact and the foal from nursing.

There are various methods of weaning — some gradual and some abrupt. Some prevent nursing but allow physical contact. Others prevent all physical contact. Still others prevent physical, visual, and vocal contact. (See chapter 2 for separation anxiety as it relates to weaning.)

It is natural for a foal to call to his dam during weaning, but if he has been properly prepared he won't panic. Although Sherlock called to Sassy, he didn't push the fence line or do anything rash to join her.

Tying is a form of mechanical restraint. Zipper has learned that being tied hard and fast to a hitch rail is nothing to fear.

MECHANICAL

Mechanical restraint involves the use of equipment to gain a physical advantage over the horse. Examples of commonly used tack in mechanical restraint are rope halters, "stud" chains, nose twitches, hobbles, and other leg restraints. All horses should be thoroughly familiar and submissive to restraint with a halter and comfortable with ropes anywhere on the body and all around the legs and hooves, and under the tail, belly, and flank.

Every horse should learn the yield response (see chapter 8) and lower his head when pressure is applied to the poll or nasal bone. This is accomplished by mechanical restraint with a rope halter and a link to manual cues, so that later, just the manual cue is necessary.

PSYCHOLOGICAL

Psychological restraint is learned obedience. For example, once a horse has learned that he can be controlled by a halter or a bridle, you can make an association between that means of control and a more subtle cue, such as a voice command or a specific body motion or position. This is how a horse can be led or ridden "tackless."

A rope halter is an ideal training aid because it can be fitted properly and applies pressure where it is needed — the poll, the bridge of the nose, and the jowls.

Besides bridleless demonstrations, psychological restraint has many valuable everyday implications. It creates a calm, submissive state of mind in the horse so that his whole manner becomes mellower. Daily management tasks such as feeding and cleaning pens are safer when a horse can be restrained by voice commands or body language. In some situations mechanical restraint (tack) is not essential as a means of absolute control.

As a horse is learning restraint, he can also learn a connection to a specific voice command, such as "easy" or "down." Getting a horse to quiet down or lower his head using your voice alone is an example of pure psychological restraint.

Subtle changes in your body position during any kind of handling, such as catching or longeing, are also examples of psychological restraint. This can be as small as a shift in weight from one foot to the other, a slight tip of your head, or a flick of a finger. Psychological restraint results in willing obedience, the ultimate in control and cooperation. Willingness is the goal of any sequence of restraint measures.

Psychological restraint results in willing obedience, the ultimate in control and cooperation.

CHEMICAL

Chemical restraint refers to drugs administered by a veterinarian to relieve anxiety or gain absolute control of a horse, such as for surgery. Some products, however, used indiscriminately, can cause unsteadiness or disorientation in the horse and make him even more difficult to control. Some

The Goal: Psychological Restraint

Ideally, training evolves from mechanical and manual restraint to psychological. For example, when you are teaching a horse to put his head down (for bridling, deworming, and so on), you'll find that his initial reflex is to raise it. When the poll is touched, the head goes up as a protective reaction. To teach a horse to lower his head instead, it may be necessary during the first few lessons to use appropriate tugs on a well-fitted rope halter or a halter chain at the same time that you apply hand pressure to the poll. Once a horse tries the option of lowering his head, you release the pressure and reward him. When a horse lowers his head and neck, he immediately becomes calmer, a further self-induced reward. Horses quickly learn to find that "happy place" and soon, neither the mechanical nor manual cues are needed — the horse responds to his handler's body language, manner, and voice. He is, in effect, being willingly psychologically restrained.

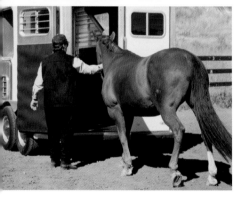

TOP: Manual and psychological restraint leads to smooth bridling. Seeker interprets the hand on her poll as a soothing signal.

MIDDLE: Mechanical (rope) and chemical (sedative and anesthetic) restraint lead to smooth castration.

BOTTOM: Psychological and manual restraint are evident when loading Zinger without any tack.

tranquilizers can make a horse appear so relaxed that handlers overestimate their safety and can be caught off guard when he suddenly "wakes up" and reacts in an explosive manner. This is especially true when tranquilizers are used in combination with mechanical restraint. You should never administer drugs unless you are qualified to do so or are acting under the direct supervision and instruction of your veterinarian.

Rules of Restraint

Since restraint is such a key element in a horse's education and because its application is potentially dangerous, keep these cardinal rules in mind.

- Never apply restraint in anger or as a means of punishment.
- Select the appropriate type of restraint for the situation.
- Practice restraint techniques before you need them.

- Make sure all of your horses have a thorough restraint program.
- Know what you are doing. If you are not experienced, get help.
- Use as little as necessary but as much as needed to get the job done.
- Always use strong tack and equipment and work in safe, sturdy facilities.

Never apply restraint in anger. The purpose of restraint is to cause the horse less distress in the long run, not as a means by which to "get even" with a horse for some previous misbehavior. It is sad to even have to state such an obvious rule, and yet some humans have egos and tempers that get the better of them, and they use restraint as a means of dominance. This is exactly the type of situation that has given restraint a bad name. Restraint, properly applied, has a calming effect on a horse and adds to his confidence, security, and safety.

Select the method of restraint suitable for the temperament of the horse and the task being attempted. Sometimes a situation calls for tactful persuasion, which can often be accomplished with just the strategic placement of hands on the horse's head or body. Other times, it might be necessary to gain the horse's attention and respect with something more definite, such as a sharp tug on the halter. Or a veterinarian might use a twitch to control a horse so that he can more safely do his work. In emergency cases, a horse may need to be immobilized either mechanically or chemically. If you choose the appropriate type of restraint for each situation, you will have better chance of success.

Practice restraint techniques before you need to use them. Many restraint techniques evoke exaggerated reactions from a horse the first time they are applied, however, by the second or third application, most horses realize that they are not going to be hurt and they become comfortable with the lesson. If you and your horse are familiar with the equipment and procedures before an emergency, restraint will most likely contribute a calming effect rather than add to the confusion. On the other hand, if a horse is already in a stressed situation and then receives his first restraint lesson, especially if it is from an inexperienced handler, it may turn into a disaster.

A calm horse with a thorough ground training program and a confident handler will not fear veterinary procedures.

The more we can help our horses overcome their fear of restraint, the less likely it is that they will overreact during a panic-prone situation.

Although hobbling isn't essential for every horse owner, it is a confidence-building tool. If you are not experienced, get assistance from someone who is. Zipper received a thorough restraint program as a yearling and lived a long life without panic, trauma, or injury.

For the safety of both you and your horse, make sure that each of your horses is comfortable with various restraints. Practice restraint techniques routinely. Once a horse has accepted the idea of restraint, a simple review of a few exercises puts him in the proper frame of mind. These can be as simple as "head down" or "move the hindquarters," or lifting up a foot with your hand or a rope.

If you are not experienced with restraint, get help with your lessons. An experienced horseman should be present at all restraint sessions. Whether taking an active part in the restraint or serving as an advisor, the experienced trainer can provide important information and feedback. For example, in order to be effective, restraint must be applied with optimal intensity — too little is ineffective and too much is unnecessary and often encourages a horse to fight.

Build a release or reward into each restraint lesson. If a horse does not receive a positive physical sign (reward, release) that he is behaving properly, he will have no reason for continued compliance now or in the future. This is important because restraint should not be thought of as a long-term training technique. It is a stepping stone — a temporary means to an end.

Link restraint with voice commands and subtle trainer body language. This will eventually teach a horse to respond to merely the voice commands and body language. In many instances the actual restraint will no longer be needed. An experienced horseman can help design restraint sessions so that they are aimed toward such future cooperation.

Safety Precautions

All restraint lessons must be carried out in safe and sturdy facilities, using strong, well-fitted equipment, and following safety practices. Small enclosures with sturdy fences or walls are preferable. Footing should be dry, level, not excessively deep, and free of large rocks that could injure a falling horse or person. Restraining a horse in a wire pen or an open area is possible but more difficult, and you run the risk of injury or escape. Loose halters, frayed ropes, or weak leather are also open invitations for failure. A horse can learn the wrong lesson if, just as he is testing the limits of the equipment, it breaks.

If you think your horse will react explosively to the application of a particular piece of equipment, you can wrap or pad his legs, knees, and head to safeguard against injury. Some trainers routinely use leg boots on their horses to offer support and

Is the Twitch a Drug?

Is the twitch purely a mechanical means of restraint or is there a chemical element too? When a twitch is applied to the nose or an ear is pinched, it is said that horses produce endorphins, which are natural opiates. The glazed-over expression of a twitched horse may reflect the release of those calming chemicals. The restraining effect of a twitch lasts less than 5 minutes, however, so the mechanics of whether and how any chemical release occurs are unknown at this time.

protection during training. You should always wear gloves and substantial footwear for protection. Keep your hair from interfering with your vision by tying it back or holding it out of the way with a hat, cap, or safety helmet. Eyeglasses or sunglasses should be well secured or temporarily stored elsewhere. Carry a pocket knife in the event that ropes or straps must be cut.

Always carefully think over the steps involved in the method to be used and discuss the plan with any assistants. Practice techniques, such as tying specific knots or applying a particular piece of equipment, well ahead of the time that you will need them. If you are proficient, your timing will be right and there will be a better chance of things going well.

Being confined in a pen is part of the life of a modern domestic horse. Even though our horses here at Long Tail Ranch spend more than half the year on pasture, when it comes time to be confined by pens they are content.

Restriction

Test A. Turn two horses out together in a large pen or pasture. After a couple of hours, separate them, returning them to their individual living quarters. There should be no calling or pacing. Both of them should stand in their pens or mosey around at a quiet walk.

Test B. Plan this test with your riding partner. While out riding on the trail, ride ahead, out of sight of your riding partner. Once you are out of each other's sight, both of you should stop your horses for 1 or 2 minutes. Then, the rider in the rear should join the rider in the front. During the halt, there should be no forward, sideways, or backward movement; no pawing, no head tossing, and no calling. Exchange places and repeat the test.

Restraint

Test A. Tie your horse to a hitch rail for ½ to 1 hour (below left). He should stand quietly without pawing, swerving, whinnying, or moving around.

Test B. With your horse haltered, exert light downward pressure on his poll with your fingers. He should lower his head (below right).

Test C. Walk into your horse's pen or stall and, without haltering him, pick up each of his feet (below left). He should not move and should willingly pick up each foot and stand on the other three in balance.

Test D. Park your horse trailer parallel to a fence, wall, or building about 4 feet away (below right). Lead your horse through the alley from both directions. Then back him through from both directions. He should remain level-headed and calm and walk in a flat-footed, 4-beat gait.

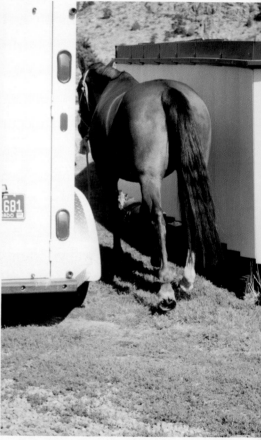

No Fear of Things

★

*Plastic grocery sacks don't kill or eat horses, but most horses
are terrified at first by their otherworldly movement and
crackling sound.*

★

HORSES ARE MASTERS AT NOTICING things, especially
when those things have changed. Because they have
excellent memories, they know when something has been
added to or is missing from a scene. They are instinctively
curious yet wary. Their highly developed senses allow them
to see, hear, smell, and feel subtleties in their surroundings.
This acuity helps them avoid danger but it can also result in
unfounded fears and sudden overreactions. For example, I've
never seen nor heard of a plastic grocery sack killing or eating a
horse, but almost every horse is initially terrified of its flapping
movement and rustling sound.

To live a safe and low-stress life, horses need to be comfortable with things in the world of humans. Although the list of potentially scary things is endless, certain key items must be introduced. As with other learning, there is carryover within and among categories, and once a horse overcomes fear of one object, he more easily overcomes fear of the next.

Sacking Out

It is best to start this process simply, with something (a stimulus) that is least likely to cause fear, gradually introduce it, and progressively increase its intensity. This approach is sometimes described as **sacking out** or gentling. The term originated in the days when cowboys used a burlap gunny sack to gentle a bronc for ranch work. In this book sacking out began when you performed the exercises in chapter 1 as you accustomed your horse to being touched all over by human hands. It continues in its traditional sense when you introduce him to tack and other items.

Horses often hear and see things that we do not. When Aria tells me that something interesting is happening down the valley, I let her look and I enjoy her intensity.

Signs of Relaxation

When a horse is secure, confident, and content, he will exhibit these signs:

- ★ Lowered head and neck
- ★ Resting a hind leg
- ★ Soft eyes
- ★ Relaxed ears
- ★ Soft lips and nostrils

- ★ Exhaling
- ★ Blowing through nostrils
- ★ Moving tongue and licking lips
- ★ Chewing

Teaching a Horse to Handle Stress

The goal of sacking out is not to scare a horse but to desensitize him to a certain degree to touch, sights, smells, and sounds while building his trust and increasing his tolerance for stress. Stress is a demand for adaptation and is necessary for growth. Think of good stress as low-level stimulation. Without stress, there is no growth. Muscles and bones must be stressed to grow strong. The same goes for the mind and psyche of your horse. A horse needs to be challenged mentally and emotionally in order to develop.

Too much stress, on the other hand, can stifle growth and development. Overload your horse's circuits and he could have a meltdown. So there's the rub. How much stimulation does it take to go from healthy stress to overstress?

TOP: Desensitization is an essential lesson. Richard gives Seeker a thorough sacking.

BOTTOM: Once a horse has no fear of odd noisy objects in his blind spot, he will be a safer mount. Zinger is on the honor system while I put on my slicker.

TOP: The horses used by mounted police are thoroughly trained and solid amid 360-degree stress.

BOTTOM: A rest break is a great reward. Both horse and rider deserve a time-out after their cross-country work.

It's actually quite amazing what horses can learn to withstand. I have been privileged to guest-ride with mounted police patrols in several large cities and participate in crowd-control exercises complete with small bombs, gunshots, smoke, vehicles, people portraying a combative crowd, noise, and commotion. Although our average riding horses probably won't need that sort of desensitization, at the moment a car backfires, your hat blows off, a pheasant flies up suddenly, or you turn the corner to see a huge flapping blue tarp sailing over a haystack, you'd like your horse to stay hinged. We need, therefore, to teach our horses to accept the sights, sounds, smells, and touch of things in our world so that they feel secure and will be safe and useful for us. We do that by introducing things in a nonthreatening way.

Build Confidence through Achievable Goals

With each successful sacking lesson, a horse's confidence grows. The key is setting small achievable goals. Your actions can and should stress the horse in incrementally larger doses, but they should never terrify him or cause him to panic. When a horse actively resists by breaking equipment, pulling away, or bolting, you have stimulated him beyond his current ability to cope. If you see that he is ready to blow up or flee, ease up and then gradually reintroduce the stress until you get back to his current tolerance level. Then, use repetition to reestablish that level. Once the horse is relaxed, reward him with a break and a rub or kind word and then increase the amount of stress.

The process of familiarizing a horse with each item varies. A saddle blanket might take minutes, but conquering a horse's nemesis, the ubiquitous plastic grocery sack, may take days or even weeks before he thoroughly accepts it.

Rewards That Horses Understand

As soon after the desirable behavior as possible (within a few seconds), reward each instance of effort in any of these ways:

- Rest break
- Walk on a loose rein
- Scratch on the withers
- Rub on the forehead
- Soothing voice

Avoid Oversacking

Although the goal of sacking out is for the horse to remain calm when confronted with spooky objects or suspicious circumstances, it is undesirable to make him so tuned out or numb that he becomes a dullard. Desensitization can be carried to extremes and produce a totally insensitive and unresponsive horse who ignores his environment and even your cues. The art of horse training is to produce an individual who is safe and sensible yet responsive.

Four Ways to Spook

A horse can react to frightening things in several ways:

1. He might **stop and freeze.**
2. He might **spook in place.** This means the horse reacts within his skin but doesn't change what he is doing — he tenses but keeps standing or keeps going.
3. He might **shy** (jump sideways or backward) and then stop.
4. He might **bolt** (shy and run) in panic.

A horse's flight mechanism is driven by self-preservation and has been part of his nature since eohippus. That's why still today, when a horse is wary of something unusual, his instincts tell him to flee (move away) rather than to stick around and fight or ask questions. Once a horse is in a state of panic, he will often forget past training, even if the associations were well-learned, positive lessons.

Sacking out and a wide range of experiences help build a horse's confidence so that when he is startled, he doesn't panic. If he has received proper sacking out lessons, he will more likely just notice something with an alert expression, perhaps pausing or stopping. At the most, he will spook in place.

Getting a horse used to a whip and rope as "regular stuff" will decrease his fear of them. Richard flips the rope toward Sherlock's head until it is no big deal (top). I stroke Seeker all over with a whip, including her topline and other ticklish areas (bottom).

The Crouching Man

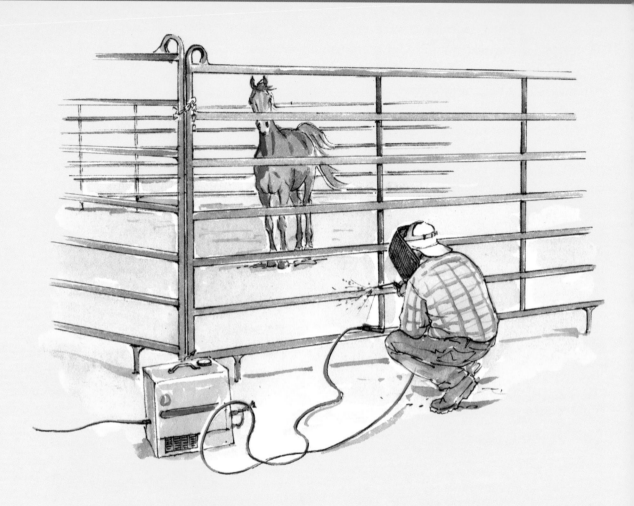

Zinger, one of my most unflappable horses, suddenly started showing an idiosyncrasy every time someone crouched down 10 or more feet away from her. For example, if she was eating peacefully in her pen and I was walking up toward the barn but stopped 20 feet away to tie my shoe, her head would come up and she would produce a long whistling alarm as she tried to get a fix on me.

By back-tracking, Richard and I figured out that several months earlier he had been crouching down, repairing the lower portion of some metal panels in a pen that was in view of her pen. The sounds and sparks from the arc welder must have frightened her, yet she wasn't the type to race around her pen so Richard, focused on his welding, didn't notice her elevated head and whistle. Her reaction diminished over time as she saw more welding, bug zappers, crouching people, and ranch activities in general, but for her entire life, she was always suspicious of the crouching man.

Things That Scare Horses

It is not just how particular things look, but how they move, don't move, sound, smell, or feel that can startle a horse. Context plays a big part too. Where and when these things appear can make all the difference.

Scary Sights

"What the heck is that?" How many times has your horse had that expression on his face? If he's not running backwards when he looks like that, it's endearing because it is so *horse*.

It's not just manmade things that can cause a horse concern. I've had plenty of experiences out in the remote mountains where my horse will suddenly see a rock or a stump or sunlight hitting some waving branches just so, and he can't quite figure out exactly what type of monster it is. But certainly, the majority of spooks come from the likes of a bicyclist, a flapping tarp, or something manmade on the ground dead ahead. Even a familiar person who appears in a hat or sunglasses for the first time can cause a horse to do a double take.

Besides wondering about a thing itself, horses react to its movement — or lack thereof. It seems as though things can be equally suspicious whether they are stationary *or* moving. For example, a parked tractor may or may not be as scary as one that is moving.

CASE STUDY: THE UMBRELLA AMBUSH

One day I was taking a dressage lesson in the outdoor arena of a friend. This farm did not have a shelter for the instructor, so to offer him some shade from the sun, the hostess appeared with a huge colorful patio umbrella. She opened it over the instructor's head about the time I was riding Zinger past him. My levelheaded horse was on the bit and on the aids so she kept fairly organized, but what an incredible welling up of energy there was underneath me! It was as though Zinger was a big balloon that was suddenly inflated to the max but, thankfully, never burst.

It took several minutes before she deflated to her pre-umbrella level. On each subsequent pass, she eyed the umbrella less and

Horses are wonderfully curious.

Visually Spooky Things

★ Plastic bags, sheets, and tarps
★ A rope moving on the ground like a snake
★ Tack (especially coming at the horse from behind, or presented from the front in the horse's blind spot)
★ Smoke
★ Machinery and equipment
★ People carrying large sheets of cardboard, metal, or paper
★ Bicycles and motorcycles
★ Umbrellas
★ Balloons
★ An unusual animal
★ People behaving strangely
★ A broom sweeping underfoot

The first time I showed an umbrella to Seeker, she was seeking . . . until I opened it, and she went "Wow" but didn't move a hoof (spooking in place) . . . and two seconds later she was again seeking.

less and finally accepted it as part of the scene. Things such as umbrellas that suddenly appear out of nowhere are especially surprising to horses.

Worrisome Touch

Once a horse has accepted the touch of our hands, it is just the beginning. To be able to groom, care for, train, ride, and use our horses, we need to touch them with all sorts of things. Some of the items we use in horsekeeping and training are scary by themselves, but when we move them or turn on their motors they become even more worrisome.

When touching a horse with any item, a firm touch is better than a tickle. For example, if you run a blanket over the tips of the hairs on a horse's back, he will most likely shudder, twitch, or shake as though a band of flies were after him. If you place a blanket firmly on a horse's back, however, he will be more likely to accept it.

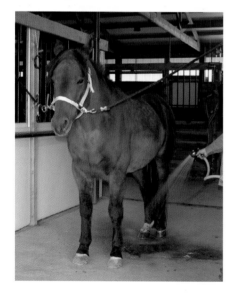

ABOVE: The on-off and spray of a hose can be startling. Seeker is an old hand.

RIGHT: Let a horse see and smell new equipment before turning it on. Dickens, meet vacuum.

PLASTIC CAN BE EVERYWHERE

There is something about plastic that is very threatening to horses. It is not just the various sounds it makes as it crackles or cracks in the wind; plastic moves in mysterious, unpredictable ways that must be very difficult for horses to resolve. But this material is a fact of our modern life, so we need to get our horses used to it in its many forms.

- Straight ahead
- Directly in the next few steps on the path
- Underfoot
- Blowing across the ground
- Caught on a fence or arena rail and whipping in the wind
- In a horse's pen where there never had been
- Smelling like gasoline
- Tied behind the cantle
- Caught on a horse's cinch and dragging under his belly
- Caught in the horse's shoe
- Caught in the horse's tail

Unusual Touch Sensations

- ★ Blankets
- ★ Whips
- ★ Plastic
- ★ Water
- ★ Slickers
- ★ Grooming tools
- ★ Hats (farrier's brushing belly, rider's blowing off onto rump)
- ★ Rope
- ★ Bridles
- ★ Bits
- ★ Saddles (strings, flapping stirrups)
- ★ Clippers
- ★ The suction of a vacuum
- ★ The weight of the rider

Plastic is one of the most fear-inspiring items to a horse because of its sound, look, and movement. Sassy steps decisively on the blue plastic even though it looks odd, sounds odd when her hoof lands, and slips around as she walks over it.

For safety's sake, every horse needs to be thoroughly familiar with ropes and all of their movements and actions. We use ropes and lines in every aspect of training: in-hand work, longeing, long lining, and riding. Whether you want your horse to be as rope broke as a rodeo or ranch horse will depend on what you will use him for. But all horses should be calm when ropes touch or even accidentally hit them on any part of the body, including the ticklish areas of the flank, head, belly, and hindquarters.

Ropes and Lines

Type of Rope		Description	Example	Purpose
Tie Rope		10–14-foot strong rope, resistant to moisture and damage from UV	Nylon or polyester	Tying at hitch rail
Lead Rope		8–14-foot thick rope, comfortable to hold, with good grip	Cotton	In-hand work
Guide Line		14–20-foot medium-diameter rope with good feel and life	Marine rope or yacht braid	Work farther from the horse
Longe Line		20–35-foot lightweight rope or webbing	Cotton web	Longeing horse in circle
Driving Lines		25–40-foot web or rope	Cotton web	Ground driving

In addition to normal rope sacking, work toward these goals to make a horse even more rope savvy:

- Swing a rope over the horse's head from the ground
- Swing a rope over the horse's head while mounted
- Swing a rope over the horse's head and let it fall humanely on his rump, neck
- Drag items behind the horse with a rope
- Drag items in front of the horse with a rope

In addition, a horse should be familiar with becoming wrapped up in ropes so that when it happens, he doesn't panic. To accustom him to this, apply ropes and pressure with ropes to the legs in various ways, under the tail, all along the belly and flank, and around the head.

BRIDLE WISE

All horses need to be thoroughly familiar and comfortable with haltering, bridling, acceptance of the bit, and being ridden with a bridle. Part of this is accomplished when you remove a horse's fear of being touched all over the head and mouth area. Once you can bridle and unbridle a horse easily, allowing a horse to wear the bridle during in-hand and longeing work will further decrease his concern about the bit and the headstall. The final piece of the bridle puzzle is teaching a horse about yielding to bit pressure and moving forward into contact (see upcoming chapters).

LEFT, TOP: Thoroughly sacking a horse with ropes is important, as ropes and lines are part of the horse life. Seeker is unconcerned.

RIGHT, TOP: Grooming requires a horse to allow all parts of his body to be touched. Sherlock enjoys getting a dust-off by Richard.

RIGHT, BOTTOM: The long lines that wrap around a horse while driving can be threatening. Casper accepts the lines resting on his gaskin-hock area as he works.

REAL WORLD
Pick Up the Duck

I used to ride Zinger in open trail classes in hunt-seat attire in the Midwest. I figured it was a great place to test our skills, and it was fun! I was often the only non-Western competitor, which could work to my advantage or disadvantage. Open shows are not regulated by any rulebook so you never know what is going to be in the course until they set up the class.

Riding in a snaffle bit on contact gave me great potential for precision obstacle work, such as side-pass or backing through an L, but when it came time to pick up or carry things, it could be more difficult for me than for my neck-reining counterparts. Never more so than the day the course included dismounting, taking a duck out of a wire cage, and then remounting, holding the duck.

Zinger noticed the duck over there in the cage as soon as we entered the course. It was probably in the back of her mind the whole time we worked the gate, went over the bridge, rotated with her front feet in a tractor tire, and loaded in the back half of a stock trailer behind squealing pigs.

When it was time to do the duck, I stopped Zinger square, gave her a rub on the withers as I dismounted, took the reins over her head, opened the cage, removed the duck, held it against my immaculate hunt-seat coat, put the reins over her head, and dropped them on her withers. She had to be on the honor system while I mounted because I had my hands full with the duck.

Now I needed to ride over to the exit gate and hand the duck to the ring steward. I had been holding the bird quite securely so at the handoff it was eager to exercise its wings. In the process, it quacked loudly and gave my saddle a little autograph with its toenails.

All the while, Zinger was very alert but a real trooper. Her senses had a workout that afternoon — a variety of visuals, several barnyard smells and sounds, flapping wings, and more. Her previous experiences with all sorts of unusual things helped to keep her grounded that day even though she had never been asked to do that particular array of tasks before.

Frightening Sounds

How sensitive a horse is to sounds will vary according to the individual. Some show horses are routinely exhibited with cotton in their ears because of their sensitivity to the noise of the announcer or the crowd. Within reason, a horse should be accustomed to the sounds he is likely to encounter.

Most routine management-related sounds, such as clippers, water spray, and the vacuum, can be gradually introduced and intensified, as described earlier in the section on sacking out. Other sounds, such as thunder or a vehicle backfire, you can't predict or easily train for. If you build your horse's confidence, however, using similar but controllable things like rustling plastic, fireworks, gunshots, or by beating a drum or bucket, you can minimize his reaction to the sudden noises of weather or vehicles.

CASE STUDY: ARIA AND THE RIDING MOWER

Early on, Aria made it known that she would rather not buddy up with the riding lawnmower. Whenever she saw it, she'd move to the farthest part of her pen. Whether it was running or turned off and parked, she'd obediently go by it when led or ridden but would tense up and tilt her head and neck in a question mark just at the moment she passed it. To help her overcome her fear, I set up a specific program that spanned about a week.

One person drove the lawnmower in the arena. The other led Aria behind, starting about 20 feet (6.0 m) away and gradually reducing the distance until Aria was herding the lawnmower. Then she was led alongside the moving mower, at first 20 feet away and slightly behind it, then gradually closer and slightly ahead of it. Gradually, Aria was relaxed walking right next to the moving machine.

The next step was driving the mower behind Aria using the same procedure. The in-hand finale was driving in circles around her as she was being led.

The entire procedure was repeated with Aria being ridden. The thorough step-by-step strategy reduced her concerns about the machine by about 95 percent, to a point where Aria felt confident and I felt safe riding her past a lawnmower in action.

Frightening Sounds

- ★ Plastic rattling, popping, cracking
- ★ Paper rustling, flapping
- ★ Clothes rustling, flapping, such as a slicker or nylon jacket
- ★ Water from a hose, hissing and spurting
- ★ Electric clippers
- ★ Vacuum
- ★ Gunshot
- ★ Vehicles' normal engine sounds, backfire, Jake brake (noisy diesel engine brake)
- ★ Thunder

Odd or Disturbing Smells

★ Alcohol

★ Medicine

★ Perfume

★ Detergents

★ Deodorizers and air fresheners

★ Smoke

★ Sheep, pigs, goats, wild game

Peculiar Smells

Certain smells can get a horse's attention, but this is a category that usually doesn't cause panic, just avoidance. Odd animal smells like pigs, game hides, skunks, and other pungent aromas might cause a horse to stop in his tracks, but not to spook.

Perfumes, medicines, and solvents can also command a horse's attention and cause an aversion, especially if these are associated with unpleasant experiences such as poorly administered medications or injections. Some horses exhibit the **flehmen response**, curling back the upper lip, when they catch a whiff of dewormer, alcohol for injection-site prep, and similar items.

If you have a specific use for a horse, such as packing wild game or herding cattle, you should systematically familiarize your horse with the smells of his job.

CASE STUDY: SASSY MEETS SOME OTHER SPECIES

When Sassy was a young mare, a friend asked me to help him sort and move some Charolais cows and their calves. We would be working in close quarters when sorting so he cautioned me about the aggressive nature of some of those mother cows.

Almost as soon as I rode into the old wooden corral, one of the cows came at us. Sassy pinned her ears flat against her head, lowered her neck, bared her teeth, and dove toward that cow. It was an extreme situation calling for extreme measures, and it worked. That cow turned tail and buried herself in the herd.

That same year when I was riding Sassy along a county road, we met our first llama. When we got closer he stretched his neck across the fence and spat at the mare. Her reaction was as if nothing at all had happened. She just stood there. That seemed to unnerve the llama and he turned and walked away.

On a number of occasions, I saw Sassy chase rogue dogs out of the pasture. If we rode by a farm or ranch driveway and a dog came out and nipped at her heels, she let me know that she'd love to turn around and give chase, which we did on many occasions. I was particularly grateful for her aggressiveness on the day that a neighbor's wolf–dog mix got loose. He was serious, not nipping playfully but biting and snapping at her hind legs. Right after she kicked at him, I wheeled her around and she snaked after him and did not back down until he was running down his driveway with his tail between his legs.

I once had a dun mare named Poco who was a real Cadillac in the arena. On a trail ride, however, we met her nemesis — a burro. I could feel Poco's rapid heartbeat and her genuine physiological fear.

Once home, I rode her to a neighbor's place where there was a wildly spotted burro. When Poco saw Lightning, she backed up so fast that we would have been in the next county if a barbed-wire fence hadn't caught her across the hamstrings.

I asked my neighbor if I could borrow and feed Lightning for a couple of weeks. I set up living quarters for the burro in a pen adjacent to my 65-foot- (20 m) diameter round pen. I put Poco in the round pen and led Lightning toward his temporary quarters.

Even though Poco could safely have stood more than 65 feet (20 m) away from us, she tried to jump out of the round pen instead. That wasn't possible, due to the 6-foot-tall (1.80 m), 2-inch by 10-inch (5×25 cm) walls, but it revealed how deep her fear was.

Once her panic subsided, I put Lightning in his pen and let him settle in with some hay. He was "ho-hum," while Poco watched him intently from far away. That evening I fed and watered her where she stood. It was quick snatches of hay the first night. By the next day, she was wearing down a little — being on red alert 24/7 was tiring. After a few days, I moved her feed trough closer to Lightning, a few feet per feeding.

It took several weeks before Poco would eat right next to their common fence. Once she reached that plateau, though, she not only overcame her fear but even became aggressive toward Lightning, trying to chase him away from his feed. It took 3 full weeks of gradual exposure for this particular mare to relax around the spotted burro.

No Fear of Things

Test A. Saddle your horse. While saddling, he should not move forward, backward, or sideways. He should neither raise or lower his head excessively nor turn around to nose or nip at you or the cinch.

Test B. While your horse is loose in his pen or stall, blanket him. He should not move at all as you throw the blanket on him and fasten all the buckles and straps.

Test C. Add one or two unusual items to the area where you groom or cross tie your horse. Then lead him in and note his reaction. He should be alert but should not stop or spook or attempt to avoid the area.

Test D. With your horse tied at his normal hitch rail or grooming area, drop a brush on the ground; drag an unplugged extension cord in front of, behind, and underneath him touching his legs; roll a muck cart or wheelbarrow in the area; sweep under his belly from one side to the other. Through all this, the horse should be calm and not move.

Test E. Put trailer boots or wraps on your horse's legs. Lead him onto an elevated platform. Have him stand for 2 minutes. Back him off the platform. He should not paw or kick.

No Fear of Restriction by People with Things

★

Start with something the horse is comfortable with.
Build on that.

★

NOW IT IS TIME TO put all fearful things to rest. Your horse should have no fear of restriction by people with things — the ordinary, everyday situations that occur as we handle and care for our horses. These composite scenarios require that we eliminate a horse's fear of being in a restricted space, with people, who have items such as tack, equipment, and horse clothing.

Then we can safely bathe, clip, shoe, trailer, deworm, and otherwise care for our horses. Any time you groom, your farrier provides hoof care, or your veterinarian performs a medical procedure, the horse's behavior reflects the thoroughness of his training.

Break It Down

It helps to break each large scene down into its component parts and determine which parts are causing problems. Then work on those portions until the horse has mastered them before attempting the whole scenario again.

If a horse is fearful of clipping, for example, you'll be most successful at helping him overcome his fear if you identify which aspect of the procedure he is reacting to and then work on it. A horse might be fine with the clipping procedure, for example, but feels nervous in cross ties. If that's the case, then cross ties must be dealt with separately and come first. Or he might be fine with clipping and cross ties but hasn't learned to live and be handled separately from other horses, so he needs lessons in confidence and independence. Or a horse might be fine with all of it for 3 or 4 minutes but then begins to get antsy, so he needs to develop patience.

Once a horse is comfortable with a composite task, I purposely add different little elements to the scene to broaden his education. When a horse stands like a statue for clipping in the cross ties, for example, and he and I are the only ones in the barn, I start to embellish. Here are some ideas to get you thinking.

While I am clipping a horse that has previously mastered the clipping lesson I might:

- Drop things on the ground near and under the horse
- Have someone lead another horse in and out of the barn
- Turn on a radio
- Have someone drive a small garden tractor next to the barn and then in the barn aisle
- Have someone vacuum a horse in the next grooming stall

No matter what you do with a horse, there are a number of individual things he needs to know. Each time you work with your horse, make a mental checklist of the components of the task. That way, if there are problems, you can determine specifically what needs to be fixed.

No Fear around People with Objects

The following visual guide shows various challenging elements for a horse in everyday management and training situations that involve "people with things." A horse should not be afraid of any aspect shown. Prior to any of these specific scenarios, however, your horse must be thoroughly comfortable with:

- ★ You
- ★ Other people, farrier, vet, helper
- ★ Being away from other horses
- ★ Being tied or cross tied
- ★ Being indoors
- ★ Standing on a mat or concrete

NO FEAR OF BEING CLIPPED

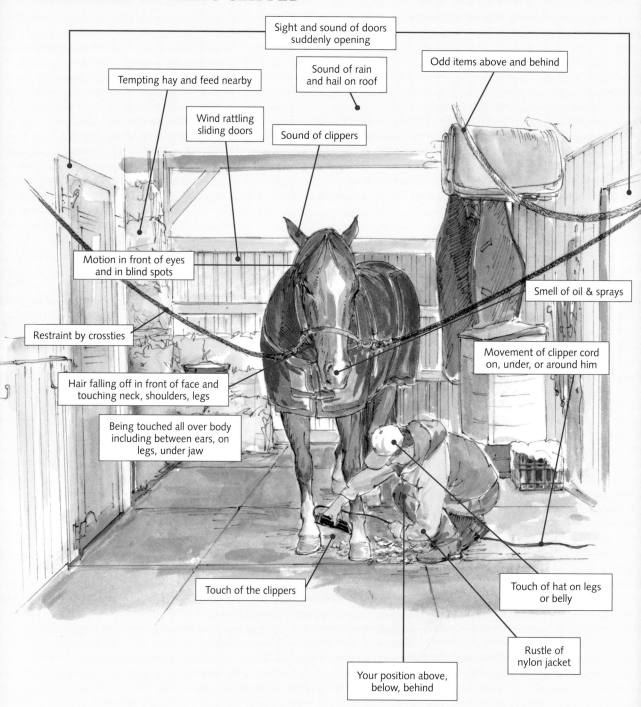

Sight and sound of doors suddenly opening

Sound of rain and hail on roof

Odd items above and behind

Tempting hay and feed nearby

Wind rattling sliding doors

Sound of clippers

Smell of oil & sprays

Motion in front of eyes and in blind spots

Movement of clipper cord on, under, or around him

Restraint by crossties

Hair falling off in front of face and touching neck, shoulders, legs

Being touched all over body including between ears, on legs, under jaw

Touch of the clippers

Touch of hat on legs or belly

Rustle of nylon jacket

Your position above, below, behind

NO FEAR OF BEING BATHED

Water on head

Being enclosed by wash rack

Sound of water spurting from hose

Restraint of cross ties

Smell of shampoos and conditioners

Your position above, below, behind blind spot

Being touched on every part of body

Water dripping from belly

Touch of hose on back of legs

Sight of hose moving on ground, often in blind spots

Touch of water, being wet on all body parts

Sight of water spraying from side and suddenly appearing in blind spot

Standing on concrete floor

NO FEAR OF BEING SHOD

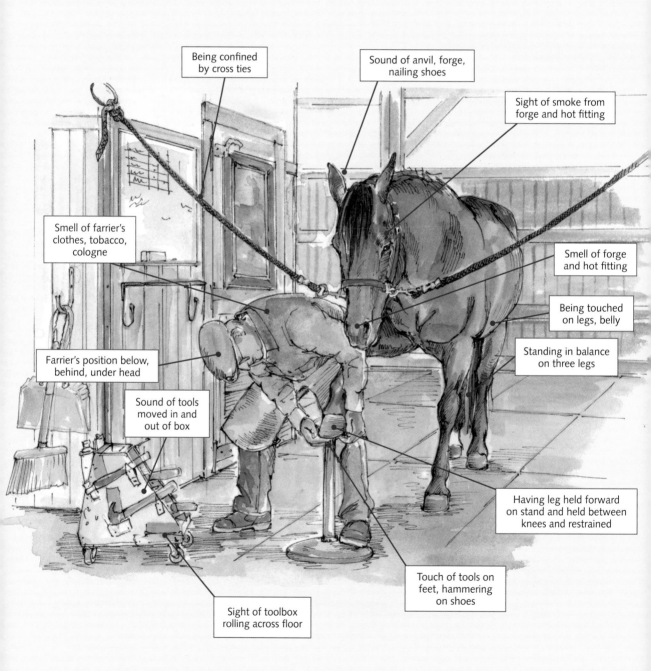

Being confined by cross ties

Sound of anvil, forge, nailing shoes

Sight of smoke from forge and hot fitting

Smell of farrier's clothes, tobacco, cologne

Smell of forge and hot fitting

Being touched on legs, belly

Farrier's position below, behind, under head

Standing in balance on three legs

Sound of tools moved in and out of box

Having leg held forward on stand and held between knees and restrained

Touch of tools on feet, hammering on shoes

Sight of toolbox rolling across floor

NO FEAR OF DEWORMING

Being touched
on head

Restraint by halter

Sight of applicator
in blind spot

Having mouth handled
and cleaned

Smell of
medication

Having medication
on tongue and in
back of throat

Presence of
applicator in mouth

NO FEAR OF VET CARE

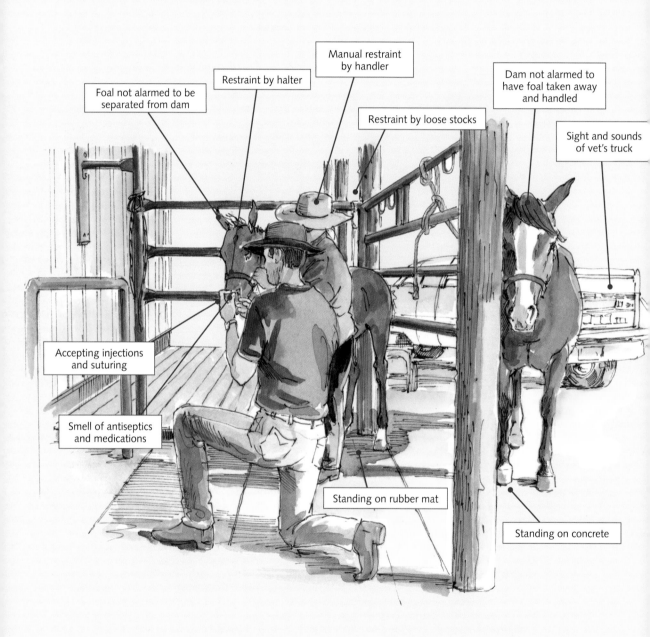

Foal not alarmed to be separated from dam

Restraint by halter

Manual restraint by handler

Restraint by loose stocks

Dam not alarmed to have foal taken away and handled

Sight and sounds of vet's truck

Accepting injections and suturing

Smell of antiseptics and medications

Standing on rubber mat

Standing on concrete

NO FEAR OF PONYING

Closeness to
another horse

Sight of person
above

Sounds of saddle creaking
and clothes rustling

Yielding to halter pressure
to come forward, slow
down, stop

Restraint by
lead line

Bending from halter
signals or position
of pony horse

Respectful distance from
pony horse and rider

NO FEAR OF BEING SADDLED

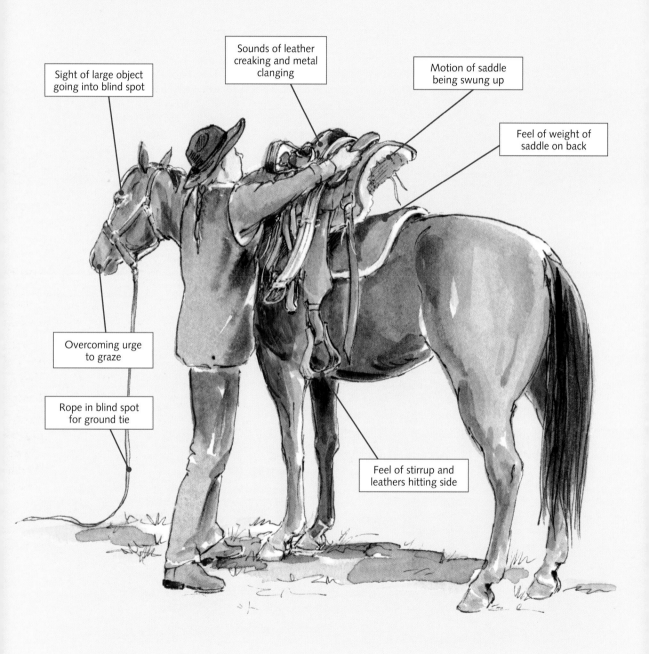

Sight of large object going into blind spot

Sounds of leather creaking and metal clanging

Motion of saddle being swung up

Feel of weight of saddle on back

Overcoming urge to graze

Rope in blind spot for ground tie

Feel of stirrup and leathers hitting side

Ready for Anything

After a full semester of working with their three-year-old horses, students in one of my college horse-training classes faced the following practical final exam. As a group, we all rode off campus to a vacant field next to an ice-cream store. The shop conveniently had a drive-up window. One by one, the students left the group and rode their young horses from the field up to the window, ordered, paid for, and received an ice-cream cone. Then, they walked back to the group in the field.

I was pleased that each horse passed this practical test. There might have been a few jog steps here and there on the return to the group, and some minor steering problems as riders rode their green horses with one hand for the first time (while holding the ice-cream cone), but the exam was designed to test adaptability. Although the students had never taken their young horses into town, and certainly not to a "ride-up" ice-cream window, they had worked the young horses individually in indoor and outdoor arenas, pastures, alongside roads, and all over the school farm, exposing the horses to a wide variety of animals, equipment, sights, and sounds. These dedicated student trainers had mentally and physically prepared their young horses to be able to deal with a brand-new situation.

PART TWO

Leadership and Partnership

By helping your horse overcome his fears, you have encouraged him to trust you. That is the first step in developing a partnership with your horse with you as the leader.

A healthy horse-human partnership is characterized by familiarity and consistency. There are no unfair surprises for your horse.

There is mutual appreciation between the partners. You provide good care and handling for your horse. Your horse provides you with respect and willingness.

You've developed an effective communication system. Both of you listen and respond and keep the lines of communication open. You never get angry and your horse doesn't get sullen.

You are the undisputed leader and you deserve that role because you act with kindness, fairness, patience, and respect.

Your horse is a willing follower, as is his nature. He is trusting, relaxed, responsive, and content.

Respect

*Mutual respect means your horse accepts you as the leader
and you appreciate him for being a horse.*

★

BEING A GOOD LEADER FOR your horse is the first step toward developing a safe and effective partnership. Because horses are born followers, it is natural for them to accept us as leaders, provided we are fair and wise. The word **leader** is really a euphemism meaning dominant one. There needs to be one leader and, in no uncertain terms, the human must be the dominant member of a human-horse relationship. A horse is much more comfortable knowing where he stands within a stable pecking order.

Horse-Human Pecking Order

Pecking order refers to the order of dominance among horses in a herd. Horses find their rank in a group through interaction with one another. Some of the factors that affect position in the pecking order are age, size, sex, physical agility and strength, and temperament. Pecking order is often established with violent kicking and biting, yet once each horse takes his place, future violence is minimal. If a horse questions his place, however, things can erupt again.

The horse-human pecking order is established through interaction as well. Because we are reasoning beings, we can usually establish our leadership without the physical violence common within horse herds.

Some humans are just not clear, consistent, and firm enough, however, to claim the top position. In those cases, horses will repeatedly test the human to see where they stand. Since it is in the horse's nature to use physical means to establish dominance, some horses and some humans will be in a constant tug of war trying to establish who is really in control.

★

It is interesting to observe an uneasy, questioning horse change into one who is calm and content, once a capable human leader shows him that his second-rung role is something he can count on.

★

Once a pecking order has been established, a group of horses can eat in peace.

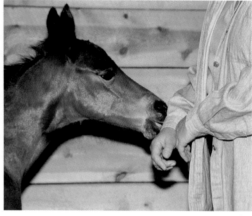

LEFT: It is natural for horses to nibble each other. These Wyoming foals are engaged in mutual grooming.

RIGHT: It is also natural for a domestic foal to try to nibble a person, but this behavior should be nipped in the bud before it becomes a dangerous habit.

When there is a capable human leader, it is very interesting to see the change from an uneasy, questioning horse to a calm, content horse, once he finds that his second-rung role is something he can count on. A horse in such a position is often more content than a horse constantly vying for position in a herd. With a trusted human leader who treats him consistently and fairly, a horse feels safe. This is the essence of mutual respect, essential for developing a partnership.

The way you see your horse will come across to him loud and clear through your body language and your manner. If you respect horses, you will appreciate them for what they are, know what makes them behave the way they do, and design your methods to train them in a way they can understand. If you do not respect horses, you might try to force them to do something or expect them to understand your intentions. Horses are horses, not humans.

On the other hand, sometimes they try to interact with humans as though we were horses too. If they are allowed to do so, it can be very confusing for them. A young foal might rear up at a human, for example, just as it would with another foal or with its mother. This is socially acceptable play among horses but should be discouraged between horses and humans.

Similarly, when a young horse is being groomed, he often wants to reciprocate as he would to his mutual grooming buddy in the pasture. His instinct might be to turn around and nibble or bite you in return. Even though such a gesture is meant to be friendly, not aggressive, intentions don't count. The best way to get a horse over this reflex is to groom him while he is tied so that he can't reach you — such as at a hitch rail or in cross ties. After a while, the impulse will subside.

Sometimes, however, even with aged horses, after a bout of satisfying mutual grooming with a horse buddy the reciprocating reflex might reappear, even if just for a few seconds. These are deeply rooted behaviors, and consistent handling is the best way to work things out.

Personal Space

Part of the mutual respect equation is acknowledging each other's personal space. Picture that you and your horse each have a "bubble" or a zone around you. At first, the horse's zone can be thought of as his comfort zone, an area with a critical distance — if you get too close, you have entered his zone and

LEFT: When a horse reaches into your personal space, remind him there is a zone he must not enter unless invited. Dickens would like to crawl into my pocket, so I remind him to not nuzzle by sending a wave through the lead rope.

RIGHT: Here there is good space and respect. Dickens is just as content staying in his zone once he learns where the boundaries are.

A horse should:

★ Know *whoa* on the guide line

★ Enter a building or a trailer (when in hand) only when you ask him to

★ Go through a gate (when in hand) only when you ask him to

A horse should never:

★ Creep up and enter your space uninvited

★ Rush past you

★ Pull away or rush away when you turn him loose

he is worried. You gradually convince him that it is safe to let you enter his zone as he lets you touch and catch him.

Once you have handled a horse and he is at ease with you, he allows and invites you into his comfort zone. When a horse stands still, he allows you into his zone. When he turns and faces you or walks toward you, he is inviting you into his zone.

Your zone also involves a critical distance. Depending on the circumstance, you will or will not want the horse to enter your bubble. If you are leading a horse through a gate or doorway, you want him to respect your space and gauge his movement so he doesn't crowd you, run over you, or swing his head over the top of your head. He needs to respect your need for space, if for nothing else but for safety. Similarly, when you are standing talking with a friend, you don't want your horse coming into the conversation by crowding or nuzzling either of you. You expect him to take advantage of the rest break and be content to just stand and wait for you.

On the other hand, when you do want your horse to come close to you, or you to him — to remove something from his eye or clip that bridle path, for example — then you want him to be well mannered when he is in your bubble. No sudden head moves, no nipping, no head butting. This is asking a lot of a horse because all of these actions are perfectly normal between horses. Nevertheless, for a horse's own safety and yours, you need to respect each other's personal space.

Unhaltering

When you turn a horse loose, you are releasing him into your personal space so he must act with respect. It is all too common that he almost squirms out of the halter as it is unbuckled and peels off running and kicking, dangerously close to the handler. A horse needs to view being turned loose as another required instance of good manners. Stand still. Wait until the halter is completely removed. Wait until the handler either gives a signal, perhaps the voice command *walk on,* or has left the area, before moving off.

One way to establish this good habit sequence is to practice it in the horse's private stall or pen. Rub the horse before and after unhaltering. Sometimes hold onto him for a few seconds with the lead rope around his neck. Sometimes drop a large wafer treat on the ground. The horse will be more interested in looking for that treat than in pulling away and running off. After a few associations with a pleasant, no-agenda lesson and perhaps a treat, you will not need to use a treat every time. The horse will make a positive connection with standing still for being turned loose.

Feeding Etiquette

Food is important to horses. If they are healthy, they eagerly await their meals. Eagerness can easily turn to aggressiveness, however, with a single horse just as it would in a herd. Prevent this aggressive behavior from starting. Even though your horse might be hungry at feeding time, he needs to have respect for you and your personal space and demonstrate excellent manners each time he is fed. At times, feeding is the only interaction a horse has with his owner, so it sets the tone for the entire relationship.

My approach is that I am a horse and that the feed is mine until I give it to another horse. I am top on the pecking order and it is my kind nature that allows me to give the feed to another horse. I am not a submissive member of the herd who lets another horse bully me into turning over the feed. If a horse tries to push his way past me to feed, he must be sent away and prevented from eating even a mouthful. If the horse were allowed to eat when being aggressive toward you, the food would be a reward for very bad behavior.

Manners at feeding time are essential for your safety. I require Blue to stay back until I allow her to approach her feed.

> *I am top of the herd pecking order and it is my kind nature that allows me to give the feed to another horse. I am not a submissive member of the herd who lets another horse bully me into turning over the feed.*

Even with a horse in a pen, it is a good idea not to dump and run. I tell Zipper to "wait" just outside his eating area while I toss his hay ration on the rubber mats. Then with an "okay," he is allowed to walk to the hay.

When I bring the feed I like my horses to:

1. Leave the eating area
2. Wait calmly outside the zone until I put out the feed
3. Have a submissive, mannerly stance when I give them the signal to . . .
4. Move into the eating area

Horses who are taught good manners individually are much more likely to act mannerly in a group as well.

Mealtime Manners

Good feeding etiquette means:

★ No biting, kicking, pawing, head shaking, threatening gestures such as bared teeth or ears flattened back
★ No rushing
★ No crowding
★ Waiting until signaled to come to the feed
★ Calm, submissive posture with head down, soft eye

Respect

Enter your horse's pen with a feed dish and a small scoop of grain.

1. Walk up to your horse and tell him *whoa.*
2. Then walk 10 feet (3 m) away from him, put the dish down, and add the grain to the dish.
3. Step away from the dish.
4. Pause for 5–10 seconds.
5. Then give your horse a cue or verbal command to allow him to come up and eat.

The horse should not move a hoof from the time you tell him *whoa* until you give him permission to move forward.

Attitude and Attention

Preserve a horse's natural curiosity, willingness to learn, good attitude, and spirit. Once lost, these precious gifts are difficult to recover.

★

To HAVE A FRUITFUL CONVERSATION with your horse, you both must be paying attention. If a horse is looking or listening elsewhere or if he is disinterested and bored, your attempts at communication will be in vain. First you will need to get his attention. Usually this is an easy task because horses are innately curious. If a lesson becomes too repetitious, however, he can soon tune out. That's why you need to find the perfect balance between consistency and variety when working with your horse.

Once you have his attention, the lessons will progress best if the horse has a willing attitude. By this I mean his overall outlook toward his work — positive or negative. While a horse might be born with a certain temperament, his attitude toward his work is formed by his experiences and his interaction with people.

Temperament vs. Attitude

I think of **temperament** as a horse's normal operating temperature. A horse is born with a certain temperament. A thin-skinned, hot-blooded mare might be naturally quick to react, very sensitive, and somewhat unreliable. Part of this comes from her biological makeup and the rest from hormones. Then, when a human gets hold of her, she can be made hotter or calmer.

On the other hand, a cold-blooded, heavy gelding might be slower to respond — less sensitive, but very steady. We can make him duller, perk him up, or turn him into a crazy horse, depending on how we interact with him.

Attitude, on the other hand, is a horse's outlook or mood when he comes to work each day. We have a lot to do with whether his attitude is positive or negative. When I was growing up and learning about horses, the phrase **workmanlike** was used to describe a conscientious horse person's attire and manner. It meant that we should arrive at the barn or show grounds tidy,

Whether a horse is a hot-blooded Arabian (above) or a cold-blooded draft horse (below) can affect his sensitivity and temperament.

alert, no nonsense, dedicated, and hard working. In a way, we are looking for some of those same things in a horse's attitude. We want him to show up ready to listen, respond, communicate, try to work with us, and make progress.

It is certainly true that temperament influences attitude. For hundreds of years, horse trainers have categorized horses by the shape of their heads, hair whorls, breed type, and many other factors. While all of these traits might have a legitimate connection to the final temperament mix, after all discussions are done, each horse is unique — he is what he is.

Some horses are naturally gregarious toward people, want to work with people, and are willing to do just about anything for us. Depending on how that type of horse is handled, he can become a partner, a pet, or a pest. I've worked with horses like that and they were real characters — on any given day, they could be a joy or a trial. Because they are like big loveable puppies, it is easy to relax the rules and let them invade your space at will. But soon they can be nuzzling your pockets and literally pushing you around.

Other horses are more standoffish and may be hesitant to bond with a human at first. I think of this type as a "horse's horse." Some of my best horses have been of this nature. On the other hand, a less social horse who is improperly handled can become difficult to catch and reluctant to go to work for you.

Get and Keep Your Horse's Attention

To have a safe, satisfying experience, know horses, know yourself, and develop a willing attitude in your horse.

Although horses respond best to a very consistent style of handling, they should not lead an existence so routine that it is boring and predictable. I've seen some horse owners who are so careful to do things the same way every single time to be consistent (which is usually good), they have inadvertently insulated their horse from the unpredictable world (which is limiting). We have a saying around our place when we see a horse get complacent or think he's a mind reader and begins anticipating: we say, "Let's mix it up."

Horses need a variety of experiences to make them well-rounded individuals. The more settings and situations they

Curiosity and Willingness

No matter what type of horse you prefer or are working with, you need to identify and preserve that horse's curiosity and interest — those are most precious assets. A willing partner is gold. One that is tuned in is unbeatable. Together you can accomplish much.

If a horse has learned to resent you or his handling, and you have to prod and remind him all day long, he might "obey" your cues, but the result won't be sterling. You might pass a particular test but you won't have a stellar team. For you to win, it is not necessary for the horse to lose. To have a safe, satisfying experience, it *is* necessary to know horses, know yourself, and develop a willing attitude in your horse.

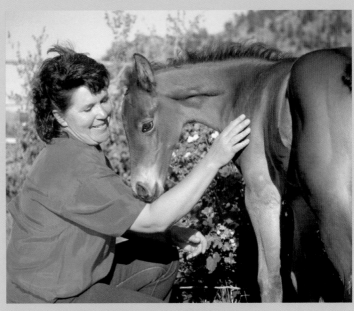

Foals are innately curious. From the time Seeker was born, I encouraged her curiosity and she has become a confident adult.

Willingness starts with that element of curiosity that horses have about people, places, and things. This curiosity should be encouraged, not thwarted. Once it is chased, drilled, or slapped away, it might never come back. Let your horse take a look. Enjoy it. That's horse.

are exposed to in a positive, nonthreatening way, the more easily they will adapt to new experiences. Variety leads to self-confidence.

Although much of the advice in this book suggests that you introduce things to your horse in a logical, progressive fashion, there are times when it is an advantage to change routines. An unpredictable menu keeps a horse keen and well-rounded and more confident about change. If nothing else, life is change. In addition, mixing it up will safeguard against boredom, tuning out, anticipation, and even sourness.

Why Is He Tuning Out?

Within reason, doing things in a different way, in a less predictable, unusual way, might help you strengthen your horse and broaden his education.

First, think about why you don't have your horse's attention. Although this is often because of a lack of respect, don't automatically assume he is saying, "I don't want to." In fact, tuning out can be caused by a myriad of physical factors: poor health, parasites, lameness, an unbalanced diet, too little or too much exercise, or other management- or training-related factors.

Lack of attention can also arise from boredom. A bored horse is often called lazy when, in fact, he is usually a keen individual. His lack of interest is characterized by a half-alert expression, little impulsion, and a delay in response to cues. If your horse is indifferent to his lessons, it is possible that you, too, are bored.

Boredom can be caused by staying in one stage of training too long, being worked in the same area day after day, or not being allowed to "be a horse." Such predictability can make a horse dull and tuned out. He has nothing to look forward to. If you vary the routine to include a new location for in-hand work, pasture or trail riding, work over ground rails, or other diversions, it will most likely pique your horse's interest.

Using a goal for the performance gives a horse (and you) purpose. Opening and closing a gate from horseback performs a useful function and teaches the horse to apply a side pass, a turn on the forehand, and a back to the negotiation of an actual

It is easier to train a horse if you have his attention. Sherlock is alert and tuned in.

object. Sorting or moving cattle, pulling a wagon or carriage, riding fence, or ponying another horse all have purpose. Participating in new sports solely for recreation — polo, hunting, mounted shooting, eventing, trail riding, or ranch-horse competitions — can be stimulating for you and your horse.

Anticipation

Anticipation is a form of tuning out because the horse isn't listening; instead, he's predicting what you are going to ask. This might be because things are repeated too often in the same manner, in the same place, over and over.

The horse who anticipates your cues might be smart, antsy, insecure, or somewhat lazier than average — or a combination of any or all! A horse who anticipates is also not paying attention, not listening to your actual aids, but rather he is going on autopilot, guessing what he expects you will be asking and acting before you actually give the cue. If he is lazy or insecure, he is looking for a quick and easy way to get through his work. He has found that memorizing a particular pattern allows him to proceed without paying close attention to you.

Paying attention requires focus and energy. When he's being ridden, a horse who anticipates might lunge into an upward transition or lurch into a turn on the hindquarters in a very disorganized and unbalanced manner. Or, he might selectively interpret a half halt as a halt and put on the brakes.

The anticipatory horse might avoid the extra concentration, both physical and mental, that is necessary to listen, communicate back and forth with his handler or rider, and work in a balanced form. His tactic is to rush through things before he can be shaped by his rider. Often, in early stages of training, an anticipatory horse is wrongly rewarded for this quickness to react because, at first, his quickness is greeted with, "Wow, he is smart; he knew what I wanted!" But later, the horse is constantly trying to outdo himself, and pretty soon that anticipation can become a slippery slope of noncommunication. It is similar to someone constantly interrupting you when you speak and finishing your sentences for you, whether it's what you meant to say or not.

To prevent boredom, try something new. Sassy and I enjoy our early morning hack.

A horse who anticipates is also not paying attention, not listening to your actual aids, but going on autopilot, guessing what he expects you will be asking and acting before you actually give the cue.

If an arena horse is showing signs of sourness, hit the trail. Zinger and I head out on an adventure and leave our dressage lessons behind us for the time being.

In other cases, horses may have forgotten — or have never learned — how to pay close attention to their trainer, so each day it seems they have to be retrained. The goal is an interactive communication and one step at a time.

Sourness

A sour horse dislikes his work and is pretty easy to identify. Signs of sourness include pinned ears, switching tail, head shaking, crow hops, bucking, lurching, and lunging. But probably the worst sour horse behavior is **balking**, which is not moving at all. (Remember that the absence of movement still constitutes a behavior.) The extreme example of balking is the horse who is rigid and unyielding and refuses to move forward, but it can also be as subtle as slowing, stiffening, or swerving.

A horse with a sour attitude might show his sourness toward a particular person, place, and activity or be sour in general. A sour horse could have been treated horribly or could have been treated fairly and just be an intolerant horse. In any case, the power of association works both ways — it reinforces bad as well as good experiences. In the case of a sour horse, he holds a grudge and acts out each time he is handled or ridden. He might also show this grouchy attitude when turned out with other horses.

When a horse has a sour attitude, it could be caused by physical factors mentioned previously, it could be related to temperament, or it could be a result of poor training and interaction with people. To understand the basis for even a mild case of balking, it is helpful to examine cases of extreme behavior disorders.

Neurotic vs. Willful

There is a difference between a neurotic horse and a willful horse. Nervous-system damage, brain tumors, and hormone imbalances can lead to neurotic behavior. There is a physiological basis for the horse's panic responses, which include a pounding heart, sweating, and panting. The neurotic horse might be physiologically frozen in fear about going forward or might wheel or spin, back up rapidly, or run sideways. A horse

Keeping Things Interesting

With fairness in mind, design your training sessions so they contain surprises for all horses, but most importantly, for a bored or anticipatory horse. For example:

★ When changing direction (change of rein) on a lazy horse, just at the point that the horse begins to fade, duck, or lean, guide him forward into a balanced, crisp halt instead.

★ If the horse is strong in the bridle, add a rein back.

★ For the horse who anticipates a canter depart, prepare him with a check (half halt) and then a halt, or work him in a large circle at a collected trot.

★ A horse who charges through lead changes should be held on the counter lead as a matter of obedience and suppling.

★ A horse who is looking for a halt or a downward transition should be checked (half halt) and then sent forward with increased energy.

If your horse is tuning out due to boredom, you can liven things up with a new skill. Sassy is clearly paying attention to the silver plastic tarp.

---

like this is usually dangerous and often untrainable except by the most astute trainers.

A willful horse, on the other hand, has probably been mentally damaged (spoiled) by previous handling and has learned that it is safer and more predictable to tune out, stand still, and take punishment than to go forward and experience the unknown. Typically, a willful horse mentally withdraws almost as though in a hypnotic trance. He may hold his breath and bite his tongue and stand with tonic immobility. The performance horse who refuses to react to his rider's cues might be choosing the seemingly safe route. The normal fussing he gets from his rider is at least predictable. It is as though a sour horse has lost the initiative to try something new because he is afraid of the unknown outcome, which, in the past, has been mostly very unpleasant. This is a variation of balking.

How Horses Grow Sour

Horses can develop a sour attitude toward their work when the lines of communication between a rider and the horse

★
The extreme example of balking is the horse who is rigid and unyielding and refuses to move forward, but it can also be as subtle as slowing, stiffening, or swerving.
★

Developing Self Carriage

Which is more humane? To nag at a horse throughout his life with kicking and strong rein aids in an attempt to hold him together, or to use a well-applied boot heel or tap with the whip once or twice to teach him to hold himself up? I've seen that constant low-level kicking dulls a horse's sides and holding a horse up with the reins leads to a dull mouth. And so ridden, such a horse will never develop self carriage. I also know horses are more comfortable knowing what is expected of them, so if you show a horse what you want him to do and you are clear and consistent, he will most likely do what you ask with a willing attitude. If you want to develop a light horse with self carriage, show him what adjustments he needs to make in his body position — don't try to hold him in position.

Self carriage is when a horse carries himself in a balanced frame without being held together on the aids. Zinger rounds into her work on light contact, even in the rain and over granite.

have been broken. You can break the communication sequence if you forget the importance of acknowledgement and yielding. If you don't yield (reward) when the horse does something correctly but, instead, you try to "hold" your horse in a particular form, the horse never learns to develop his own natural movement, balance, and cadence. (For more on yielding, see chapter 8.)

Of course, the line of communication can also be disrupted by the horse if he doesn't answer at all or answers incorrectly. In this case, you would need to use repetition of the aids with increased intensity.

Sometimes a horse gets mentally overstressed or physically sore, and then becomes sour because his training has been rushed. A "30-day wonder" job on a three-year-old often doesn't give the horse sufficient time to adjust to the pressures on his mind and body. Initially, a mentally or physically confused horse will move his ears back and forth as he tries to figure things out. Later, his wondering could turn into resistance and he puts his ears permanently back (closed off) and tunes out.

Some sour horses simply do not like their work or are not suited for a particular event, but are being forced to perform in

a difficult and possibly painful manner for their conformation. Pinching or bruising at the withers, back, tongue, or lips from ill-fitting tack or incompetent riding can also cause extreme and long-standing sourness. A simple inspection of the bearing surface of a saddle might reveal the cause for the horse's discomfort and sourness. A well-balanced Western saddle distributes the rider's weight over a larger surface area than most English saddles. Using a Western saddle temporarily on a young or weakly muscled horse may make a positive association with riding until the back becomes conditioned and better able to tolerate an English saddle.

Even a seasoned horse with a good rider but inadequate conditioning can experience muscular and skeletal soreness that might lead to sourness. A horse with an intermittent or early stage of lameness, such as navicular syndrome, or a horse with hoof cracks or tender soles will often be grouchy even if he is well trained and well ridden.

A horse who is worked too hard or too long for his level of cardiopulmonary conditioning, has had "the wind knocked out of his sail" and could be reluctant to repeat such an all-out effort with his heart and lungs again. He could likely return to work with a sour attitude.

What You Can Do

Fortunately, there are many ways to avoid and deal with boredom, anticipation, and sourness. In horse training especially, an ounce of prevention is worth a pound of cure. Once a bad habit is deeply ingrained, it is harder to change, and that's when harsh methods might be attempted. But that would be

The saddle and bridle and any other tack should be properly fitted. By the time a horse is mounted up for the first time, he should be so used to riding tack that it is like a second skin — no rubbing, no uneven pressure, no pinching.

★ *Be thankful if a horse is still acting out. A moderately spoiled horse who still acts and reacts is easier to change than a mildly spoiled horse who is depressed, tuned out, and no longer interacts with you.*

★

Spicing Up a Sour Horse

Finding alternatives to riding might brighten up a sour horse. These include longeing, longeing with a rider, in-hand obstacle work, cavalletti work on the longe, ponying, long-reining, or driving him with a cart or wagon.

not only dangerous but also, most probably, ineffective or even counterproductive.

Be thankful if a horse is still acting out. A moderately spoiled horse who still acts and reacts is easier to change than a mildly spoiled horse who is depressed, tuned out, and no longer interacts with you.

With a mild case of boredom, you can bait the horse and bring things to a head. For example, you can eliminate your use of supporting aids so that when a horse cheats, you can catch him in the act and apply appropriate aids. If your horse tends to fall lazily into the circle with his forehand, and you've

Preserve a horse's good opinion of himself by giving him work he enjoys. Zipper looks forward to trail rides almost as much as I do.

been holding up his inside shoulder with your inside leg but are getting tired of supporting him, try this:

- Relax your inner leg (bait the horse)
- Let the horse fall to the inside
- Give him a more forceful reminder (heel of your boot, tap with the whip) to get him to hold himself up

The goal is to get the horse's attention so that he reacts to, rather than leans on, your aids. Lighten the aids, let the horse make a mistake, correct it clearly and crisply, and then resume riding with quieter aids as though nothing has happened.

Rx for Attention Issues

Here are some ways to get the attention of the anticipatory or sour riding horse. Groom, tack up, mount, dismount, and put the horse away. For a bored horse, bypass an extensive warm-up and get right to the more demanding work. Tack up rather quickly and work intensively for 5 or 10 minutes, then groom the horse thoroughly before you put him away.

In contrast to the human 5-day work ethic, I've seen many horses often make more progress in their training when they get one or two lessons per week. This is especially true with young horses. If a horse has a severe attitude problem, 2 or more months off for grazing and relaxation might give you a fresh mental slate to work with when you bring the horse back to work, especially if you have evaluated and refreshed your training program in the interim. You might get further ahead in 1 week than you would have if you had kept the horse in work for those 2 months.

Above all, preserve a horse's good opinion of himself. Find an activity the horse is suited for and enjoys. Allow him to feel accomplished and confident. Remember to use positive reinforcement to let him know when he has done well. In this way, you will keep his attention and his good attitude.

A positive attitude on your part goes a long way into improving a horse's attitude. Dickens, a.k.a. Mr. Personality, makes the job easy.

Attitude and Attention

With halter and lead rope in hand, head toward your horse's pasture gate. As soon as he sees you, he should start heading for the gate, meet you there, and stand quietly while you halter him.

7

Patience

★

Sometimes the slower you go, the faster you get there.

★

ONE OF THE MOST COMMON missing links in a horse's training is patience. And that's too bad because the more a horse is trained to be patient about certain things, the more carryover there is to other activities and the calmer he becomes overall. That's why many of the exercises in ground training or riding programs emphasize stopping and standing. Patience includes slowing down, stopping, and standing still, as well as performing things in a controlled, measured way, one step at a time. All downward transitions are a combination of Restraint and Patience.

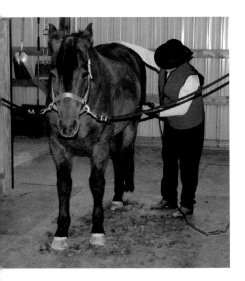

Patience is a virtue, and never more appreciated than when body clipping. Without moving a foot, Seeker dozes and allows me to take 2 hours to do a thorough job.

If you are impatient, your horse will know and it will be harder for him to develop patience. Awareness of your tendencies and body language will help you become more patient together — just one way horses can make us better people.

A Valuable Virtue

My aging part-Warmblood mares, Aria and Seeker, show signs of having equine metabolic syndrome (EMS). One of the characteristics of EMS is a very long winter coat and a thick summer coat. To keep them cool in summer, therefore, and so that they can work without sweating excessively, it seems I am always clipping them. It's not just the usual bridle path and legs session, but full body clips followed by frequent tidying up throughout the season.

I had just come in from a late-summer tune-up on Seeker and realized to my surprise that I had been clipping for almost 2 hours. During that entire time, Seeker kept very still unless I repositioned her for better access to a particular body part. Her patience made it much easier for me to clip all those nooks and crannies. In fact, while I was clipping, I thought how different that session would have been with a horse who was impatient, antsy, or frightened of clipping.

An Important Basic

It's easy to see how the Patience step can be skipped as many horse owners are eager to climb on and go do something with their horse. In some instances, such as barrel racing and other events, the main focus is "go, go, go," and it would be easy to forget the importance of stopping and just standing still. But any good trainer knows the value of just sitting on a horse and hanging out. I like to take the opportunity to view the beautiful scenery while giving my horse a break, and I expect my horse to stand there without moving or fidgeting.

From the beginning, therefore, and throughout any horse's training, it is important to devise specific exercises that develop patience.

Slowing and Stopping

To stop a horse's forward motion might be difficult if you have an untrained, fresh, and energetic horse. Eventually, the horse will need to be stopped when he is free in a pasture or pen, being led in hand, and being longed, driven, or ridden. But how you establish Whoa (Stop) and Stand will depend on the age of the horse and the stage of the horse's training.

In reality, you need to teach a horse go before you can teach him Whoa. Most horses naturally want to go, so in the early stages, we need to teach them to slow down and stop. But if your horse lacks impulsion, you should first read the sections in chapter 9 on Forward Motion.

First Stop

The first stop is often accomplished with the horse free in a pen — traditionally a round pen. In some ways, it makes sense to work a horse until he is tired enough to want to stop before you ask for a stop.

But to work smarter, rather than harder, you can use body language and timing to drive, turn, and guide a horse so that in much less time, and with much less sweat and stress, he begins to look to you for guidance. Then, something clicks in his mind and he will usually turn his head or his head and forehand toward you and stop. From that moment forward, you can build on that connection and increase the length of time you require the horse to remain still, where and when you ask him to Stop and Stand, and what type of distractions you will introduce to test his mastery of these lessons.

HOW SHOULD A HORSE STOP?

Whether you want your horse to stop on a straight line in the direction he was headed or turn and face you or even walk up to you is a matter of personal preference. I want my horses to know every variation. But at least 85 percent of the time, I want a horse to stop straight, on the track he was headed, perhaps turning only an ear or his head toward me, but not to take even one step toward me. This makes it easier to continue with the work. At other times, however, I do want my horses to stop, turn toward me, and walk up to me. In my experience, a horse really wants to do these last two things, so I limit how often I

There are times when you want a horse to stop and times you want him to stop and turn his head toward you. Long yearling Sherlock stops straight on the rail, ready to move forward at a moment's notice (top). When I want him to turn his head toward me (bottom), I scratch my nails on my jeans. I just might be asking for an inside turn next . . . good boy.

Standing square on a long line is the basis for good manners. Dickens would much rather be up close and nuzzling but he has learned this important skill well.

allow him to do them. I reserve about 15 percent of the stops for the "come to me" exercises.

This is somewhat similar to my thoughts about not over-doing backing. The more forward a horse moves, the better he will be at backing. It is not necessary — and can often be counterproductive — to do too much backing. A horse can learn to use backing as a sort of evasion when he doesn't want to accept contact with the bit or cross an obstacle. So I save backing for when I need backing.

Similarly, if every time you ask a horse to stop when longeing you teach him to turn inward and walk up to you, he will tend to look to the center as he works, wanting the stop break. So, if you see a horse developing that sort of tendency, perhaps you should emphasize that he stop straight on the track, as I do. On the other hand, if you are working with a horse that is reluctant to come to you, you might want to incorporate more "come to me" for rubbing in your training sessions.

The more forward a horse moves, the better he will be at backing.

HOW LONG SHOULD HE STAND?

As you fine-tune stopping, and perfect it from a walk, trot, and canter, you will also add how long you expect a horse to stand after the stop. Here again, you want to vary the routine. Sometimes you might want him to stop for only a few seconds and then trot off. You almost preserve some of his momentum when you ask him to stop for a very short time, so this is very good for developing forward motion and crisp halts and departs.

Other times, you might want your horse to stand for 10 seconds or even longer. Ten seconds is actually very short, but when a horse has been working and you then ask him to stop and not move a foot and stand there on the rail of the round pen or arena, it can be a challenge!

Standing Quietly

Ideally, you should be able to stop your horse and have him stand still anytime, anywhere, and for any reasonable amount of time.

In Hand

If you are walking your horse around a show grounds and stop to chat with a friend, you want your horse to stand there without fussing. Similarly, if you need to reach inside the tack room of your trailer for something, you should be able to expect your horse to stop and stand on the honor system while you reach in and get what you need. Throughout a horse's in-hand training, you should teach him to stand for longer periods of time, on progressively longer lines, and with various things going on to test his patience.

Guide Line

A horse should know Whoa on the guide line with you at the other end of a 15- to 20-foot (4.5–6.0 m) rope. The rules of Whoa on the guide line are simple: *whoa* until I say go. You should be able to walk all around your horse, on both sides, in front and at the rear, without him moving.

Ground Tie

Once he has mastered Whoa on the guide line, it will be a natural segue to Ground Tie. During Ground Tie, you drop either the lead rope of the halter or the reins of the bridle on the ground in front of the horse and tell him *whoa*. The horse should stand there until you come back to his head to lead him away.

You should be able to stop your horse anytime, anywhere, and he should stand until you ask him to do something else. Zipper waits patiently while I give directions to "that other trail over there, behind that ridge and over the draw, but before you get to the grove of aspen that is just to the north of that pond that might be dry this time of year but. . . ."

Ground Tie describes a horse that stands square without moving or grazing. A horse should ground-tie on the honor system: that is, not held by rein or rope but through respectful obedience. Although Sassy is watching something down the valley, she stands still while I tie my slicker behind my saddle.

Asking a horse to Ground Tie before he is mentally ready, however, can result in a setback. I've seen horses lose a lot of progress in their training when they follow their handler and step on the lead rope or reins as they walk away, causing a severe jerk on their head or mouth and a panic reaction. Make sure your horse has mastered Whoa on the guide line before you attempt Ground Tie.

Patience for Procedures

Hoof trimming and shoeing, floating teeth, clipping, bathing, soaking an abscessed foot in a bucket of water, holding a leg up for a flexion test — all of these things take more than a few minutes, yet many horses haven't learned to be patient enough for long enough. The more thorough your horse's No Fear and Whoa training was, the more relaxed he will be for these longer procedures.

Riding

Nowhere is patience more evident in the riding horse than during mounting. First and foremost, for your safety, your horse needs to know this is one time not to be impatient or goof around. Much should be made of the early mounting lessons. You should partially mount, then slide down, mount from both sides, and do these things over and over until your clambering over his back becomes very ho-hum. Then, once all the sensations and reactions have disappeared and your horse is patient with all of it, you can mount. Whether you dismount and remount again right away depends on your goal.

In all cases, it is best to just sit still once you do mount. Don't ask your horse to immediately dash off as soon as you mount. First, establish his habit of patience. You can always add that movie flair another day. Until your horse has impeccable mounting manners, keep working on it.

Developing Patience While Tied

Whether a horse has developed patience will really show up when he is tied. At a hitch rail , at a trailer, in cross ties **B**, alone or alongside other horses **C**, a horse should stand without swerving from side to side, pawing, bobbing his head, chewing ropes or rail, or generally fussing around. But the great thing about the lesson of tying is that besides exhibiting whether a horse has patience, it also *develops* patience. Teaching a horse that he has to stand still is one of the best tools you have. If a horse learns to stand tacked up for several hours at a time, it does wonders for his tolerance, acceptance, and patience.

First, you must be sure that the area in which you tie is suitable. The ideal location is somewhere with protection from the sun, precipitation, and flies. There should be an unbreakable, very strong place to tie, such as a post or tall hitch rail. The ground or floor should be comfortable and durable. My favorite is a heavy rubber mat over either concrete or dirt. The rubber provides somewhat of a cushion and comfort for the horse. In addition, it prevents a pawing horse from damaging his hooves, shoes, or the ground. If you set up a horse to succeed by providing him with a good place to stand, he will look forward to the rest periods and immediately kick into doze mode when you tie him up whether alone or with other horses.

Once a horse has learned to stand tied in loose stocks or at a hitch rail, he will more likely stand contentedly in cross ties and a trailer.

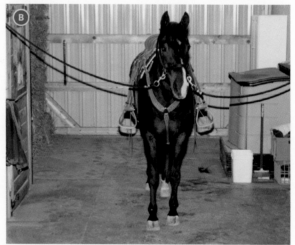

Patience refers to more than standing still. A horse must be patient when performing any maneuver or negotiating any obstacle. When you are riding, there will be countless situations where you will be glad your horse has developed patience: standing next to a gate while you work the latch; reaching over to get something out of the mailbox; waiting for your turn to go through a narrow passage on a trail or to cross a narrow creek. But you'll also want your horse to know in no uncertain terms that, when asked, he is to move "one step at a time." This is essential whether you are perfecting a sidepass or caught in a log jam.

If you are an impatient person, your horse will pick up on that and it will be harder for him to develop patience. If you are aware of yourself and your body language, you and your horse can work together to both become more patient. This is just one more way that horses can make us better people.

ABOVE: When you need a little extra time for mounting, a patient horse is greatly appreciated. Zinger stands as still as a statue for Richard to mount for his first ride after knee surgery.

RIGHT: If you are an impatient or nervous person, your horse will pick up on that. Zinger would like to see what is going on behind her, but because I am not twisting around myself and instead am focusing on what I am doing, she stays planted.

Patience

With your horse haltered and with a 20-foot (6.0 m) guide line attached, tell your horse and leave him while you get something out of your horse trailer. Make him stand for 3 minutes. He should lower his head and rest (but not graze) and not move a hoof.

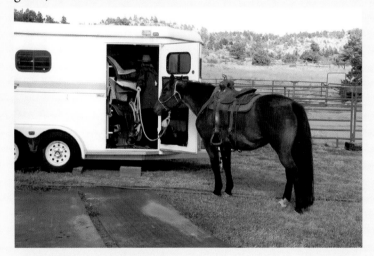

Ground Tie

Go out to your horse on pasture, halter and ground-tie him. Pick up each of his feet and clean out the hooves. He should stand perfectly still and balanced without lowering or raising his head or fussing with his feet.

8

Yielding

*Use reward and yielding to reinforce desirable behavior
and maintain a horse's good opinion of himself.*

★

I F YOU'VE DONE YOUR HOMEWORK thus far and have a horse who is respectful, patient, paying attention, and not afraid, then he already has the tools he needs for yielding and suppleness. It will just be a matter of establishing a means of communication between the two of you, because yielding is indeed a two-way conversation.

It all begins with contact. Contact was introduced in the No Fear lessons that helped your horse overcome his natural skepticism of people, things, and pressure. Horses seem to like body-to-body contact and that makes our task easier.

Where Yielding Fits In

The basic sequence of lessons goes as follows:

1. No Fear of Contact
2. Yield to Pressure
3. Move Forward into Contact

Everything that follows is based on this essential string of events. By establishing things in this order, you are more likely to end up with a fluid, supple horse who is an extension of your aids. You are more likely to become one with your horse and work in harmony.

Yielding is used in two ways in horse training. We ask our horses to yield to pressure, which is the subject of this chapter. When a horse does what we ask, we reward him, which often takes the form of a yield or release of pressure.

Reward

A reward is something the horse inherently perceives as pleasant. A horse should be rewarded when he has done the right thing. A reward does not necessarily mean a treat. In fact, food rewards, although they do have their place, can complicate things. Some trainers successfully use treats in training, but there are other options.

Inherent Rewards

With inherent rewards the horse does not have to learn they are pleasant.

A rest break. From a horse's viewpoint, a rest break after a session of loping or a particularly nice half pass is a reward. After a good effort, if you let a horse stop or mosey on a loose rein, stretch, and fill up on air, he is getting a physical reward he understands.

A good rub. If you know where and how your horse likes to be rubbed, you can give him a rub as a reward. After a nice series of longeing maneuvers, if you stop your horse and go up to him and rub him on his forehead, neck, or withers, he will most likely lower his head and exhale, relax, and feel content.

Verbal praise. A horse can associate our voice with positive reward so that in certain situations where you can't easily offer a physical reward, you can assure him he is doing just what you had hoped for by saying "Good boy."

A reward does not necessarily mean a treat. In fact, food rewards, although they do have their place, can complicate things.

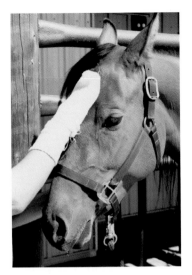

Rubbing on the forehead is a reward that almost all horses inherently appreciate. Seeker carries over the association from grooming to training.

Yielding Pressure

In a more subtle but very important way, whether you are ground training or riding, when you yield pressure on the horse, you are rewarding him.

Rein release. Release of rein pressure tells the horse he is doing well and encourages him to relax and perhaps start developing self carriage.

With rein aids, the pressure you apply to ask a horse to bend or stop should always immediately be followed by a yield when he complies. Even if his attempt is not perfect, if he tries, he should get some release to show he is on the right track.

In terms of ground training, if a horse has learned how to lead and maneuver like a butterfly on a string, you can relax your body language and your lead rope and halter cues. This tells the horse he is carrying himself and responding to you as you wish, following your natural body movements.

Body language. Release of pressure also refers to your body position and body language when you are doing in-hand work, longeing, or ground driving. Backing off, stepping to the side, looking down, and changing your bearing overall are all ways of taking the pressure off a horse.

If you don't reward through yielding, you kill the horse's incentive to comply and it will be harder for him to develop self carriage.

Encouraging Self Carriage

Self carriage means good posture during movement without being held or prodded into position. It is a combination of balance, poise, fitness, and neuromuscular conditioning. When the aids are removed (rein and leg aids passive), the horse holds himself, on his own, in a balanced form while moving.

The conformation of some horses allows them to exhibit self carriage almost naturally. Other horses can develop self carriage through training and conditioning. The goals of conditioning are an elevated forehand, a raised back, strong abdominal muscles, and engaged hindquarters.

The Yielding Conversation

Yielding by the horse is a giving to pressure applied by the human. That pressure can be as subtle as your overall manner or a shift in your weight or body position and can escalate as a step toward the horse, a hand on the horse, and signals using a variety of tack. The variations are many but the principles are the same.

Beginning Steps

In its simplest form, the yield consists of three steps:

- You ask the horse to give by applying and holding pressure.
- The horse yields.
- You release pressure (reward).

At first, even the slightest hint of a yield by a horse should be acknowledged with a reward or release. It is better to reward the horse each time he tries, even if he doesn't fully yield in the way you hope he ultimately will. You will reward successive approximations to the final desired behavior and eventually your horse will get there. That is shaping your horse — you are indeed a sculptor.

TOP LEFT AND RIGHT: A horse needs to learn to yield to halter pressure by moving forward when asked, either one step at a time or all at once. I reel in Seeker with light halter pressure.

BOTTOM: Moving over at the hitch rail (turn on the forehand) comes in handy when grooming, tacking, and shoeing. Seeker willingly complies with a deep crossover step.

Adding Refinement

Later on, in more advanced training, such as when the horse is mastering collection, vertical flexion, and self carriage, the reward is paramount. Pressure should never be continuous in an attempt to physically hold a horse in position. There should always be yielding.

As things get more advanced, the conversation becomes more refined and rich.

- You prepare the horse, such as with a precue, body language, a check (half halt).
- Your horse pays attention and is ready.
- You apply the aids.
- Your horse responds and performs (we'll assume properly).
- You reward the horse by yielding in some way.
- Your horse continues what he is doing and relaxes into it.
- You follow the movements of the horse until . . .
- A new conversation begins and you ask for the next change.

Without being asked, a horse will demonstrate willing compliance and yielding in many ways. Seeker automatically lowers her head when she feels the haltering process, which makes it easier for me because she is actually quite tall.

Sequence of Lessons

Yielding lessons begin early in the handling of a horse and are an integral part of his advanced training. Here are some examples:

- Head down
- Move over laterally
- Move over while tied
- Bend laterally
- Bend vertically
- Step back in hand
- Back when tied
- Back when ridden

It is better to reward the horse each time he tries, even if he doesn't fully yield in the way you hope he ultimately will. Eventually he will get there. That is shaping your horse — you are indeed a sculptor!

Choose Your Backing Method

Backing (rein-back) is a form of yielding based on forward movement. A horse that doesn't have good forward movement often stalls or shuffles when backing. A horse with brisk forward movement usually marches backward in a crisp, two-beat rhythm with diagonal pairs of legs lifting and landing in unison.

You'll need to back your horse out of a trailer, while tied, while opening a gate, and in other situations so it's best to have a variety of backing methods in your repertoire. Standing in front of your horse and wiggling the rope **A** as you walk toward him is a good starting point. Using light backward pressure on the noseband of the halter **B** asks for flexion at the poll. Applying light pressure at the point of the shoulder **C** is ideal when asking for one step back while tied. To test a horse's self carriage and straightness while backing, back the horse without tack **D**.

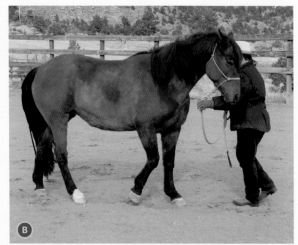

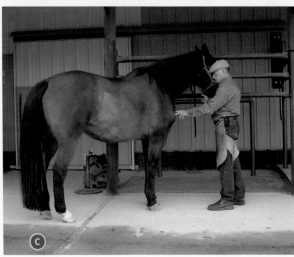

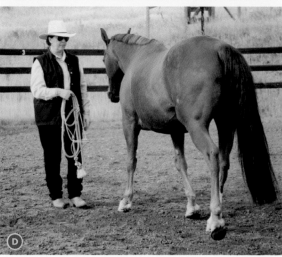

Yielding

With your horse haltered, stand alongside the girth area and draw his head around to you with the lead rope. His neck should bend and his head should come easily and fluidly but stay vertically straight. If you release pressure on the lead rope, his head should stay momentarily bent toward you and then gradually return to the straight-ahead position.

If you overbend your horse, it will be anatomically impossible for him to keep his head vertical. Although Sherlock is very supple, bending this far is only for a test but not for practical use.

Backing

Lead your horse up onto an elevated platform. Stop the horse when all four hooves are on the platform. Pause for a minute. Then back the horse off the platform. The horse should stay in the center of the platform and step off the end, not the side.

REAL WORLD
Rattlesnake!

Day 1: SHOCK DAY

At 7:30 AM, Richard and I turned Seeker and Sherlock out for a half-day graze in the southwest pasture. At noon, Sherlock headed toward the gate as usual but Seeker was "stuck" in one spot. As I approached, I saw that her left front leg was very swollen and her lip was drooping. Her normally perky personality was drained yet she gave me one of her deep whickers, which tugged at the core of my heart. At that moment her welfare became the focal point of my life. There were several drops of blood on the outside of her left pastern below the coronary band, and they spelled rattlesnake bite. The best way to get her to the barn was to lead her slowly and quietly down the hill. While Richard led her, I ran to the barn and got out the bute and the hose.

In the cool of the wash rack, I gave her the bute, offered her water, which she refused, and started cold-hosing her leg, intermittently spraying cold water onto her lips and into her mouth. That got her tongue and lips moving and began restoring her droopy lip to a more normal state.

Then, while Richard manned the hose, I ran down to the house to call the vet. He said good on the bute and to continue cold hosing. After 20 minutes of hosing, I dried her leg and clipped the area, where I found four fang marks. By this time her left cannon, knee, forearm, arm, and chest were as hard as a melon and three times normal size. I massaged her leg for about 20 minutes and hosed again. By that time the vet arrived, less than an hour from my call — something for which I am forever grateful. He concurred that it was likely a rattlesnake bite.

He inserted an IV into Seeker's jugular vein and administered one liter of saline solution to encourage her to drink (a hypertonic solution usually activates the thirst reflex). He also gave her more anti-inflammatory drugs and pain relievers by mouth. Then, intramuscularly (IM), he gave her a double dose of penicillin. We returned Seeker to her pen, and she began drinking almost immediately. I think I took my first deep breath about then.

Treatment Plan

His directions for her care included:

* Bute twice a day for 3–4 days

* Penicillin twice a day for 7 days

* Soaking in Epsom salts at a temperature as hot as my hand could tolerate

* Hot Epsom salts compresses on the forearm and knee

* Alternating with cold hosing as I saw fit (he said, "It's an art")

* Massaging the affected leg and chest as much as possible

His prognosis was guarded so I activated my three Ds — Determined, Dedicated, and Diligent. He left me with supplies to get started and prescriptions. Seeker was barely able to lift her foot an inch off the ground.

The swelling, massive now in her forearm and elbow, created a sharp shelf of tissue with a deep crease all around her forearm and into her chest. It was very painful and could even rupture, so I needed to get the swelling down quickly. I alternated cold hosing and massage. We'd used up the last of our Epsom salt on the first soak; each soak requires 2 to 4 cups. So Richard drove to the closest town with a grocery store and wiped out its Epsom-salt supply.

As I sat on a stool massaging Seeker's left leg, my head resting gently on her forearm, I devised the SCRAM plan: Soak, Compress, Rinse, And Massage.

To soak the hoof and lower leg, Richard would lift Seeker's hoof one inch as I quickly slipped a bucket through that space, so we could place her foot and the bucket down at the same time. Then I'd add the Epsom-salt solution and she'd soak 20 minutes.

Meanwhile, I mixed another, stronger Epsom potion for compresses, which I wrapped around her knee, forearm, and chest floor. I'd hold them in place and gently massage but it was very painful for her. She kept her left foot solidly in the bucket but occasionally she'd bob her head and tap the toe of her right hind shoe on the wash rack floor to tell me it hurt.

After the soak and compress, I'd rinse all the salt off Seeker's leg with cold water, buff it dry, and massage as much as possible. It seemed to soothe her when I gently scratched her leg. Another technique I used was to grab her entire forearm with both of my hands like a football and gently rock my hands back and forth to stimulate the skin layers.

After a SCRAM, which took 30–45 minutes, I'd turn Seeker back into her pen.

Thinking she would want to lie down, we bedded her adjoining stall deeply with shavings and added several tractor buckets of pea gravel to her turnout pen. But she chose to stand in her favorite place, the rubber matted eating area of her sheltered pen where she seemed to feel safe and content.

Every 2 hours, another SCRAM, and shortly after there would be droplets of serum and blood coming out of the fang marks.

At 12-hour intervals I gave her bute orally and penicillin IM. Penicillin was injected with a 1½"-long 18-gauge needle, rotating sites between the two sides of the neck and the hamstrings to minimize soreness.

The Rest of Phase 1: More SCRAMs and Dry Rubs

For the rest of the first phase, it was more of the same. Between the SCRAMs I did dry rubs, a combination of rubbing her leg, chest, and midline with a dry terry cloth or grooming gloves (the type with rubber bumps on them).

On Day 2, I tested Seeker's ability to walk to help reduce the swelling but decided it was too early.

Day 3 started full of hope, with a bright-eyed, alert mare wanting breakfast and moving around her pen for the first time on her own. SCRAMs, meds, and dry rubs all day. Early in the afternoon, I noticed dried serum on the inside of her cannon and forearm — a positive sign that the compresses had pulled out some of the swelling. So before her dry rubs, I gently washed away the sticky serum with lukewarm water and Nolvasan shampoo. Then I took her for a short walk.

Early on Day 4, if I didn't look at her leg, I'd say she was acting pretty normal

as she walked her pre-feeding laps around the pen. The swelling was now cool to the touch and softening. I wanted to make a gradual transition from SCRAMs and rubs to exercise. So it was meds as usual, and one SCRAM and three dry rubs, but also three 4-minute walks with rubs afterwards. At evening meds, I reduced the bute by half.

On Day 5, Seeker was bright and there was further reduction in swelling, but it persisted in her left front knee and forearm. The main positive change was an evenness in her gait and weight bearing. I observed her even resting a left hind, evidence that she was bearing more weight on her left front. I eliminated the bute and conferred with my vet. He suggested I switch her over to oral antibiotics (sulfa) and keep her on them for another week, twice a day.

I hand-walked her for 12 minutes and then sent her out on the 20-foot lead rope so I could watch her move. She looked really good. A few hours later, Richard took her for a 15-minute walk while I was getting lunch ready. Midafternoon, I walked her again and gave her a good dry rub.

Phase 2: Days 6–12

During the second phase, as I increased exercise to include hand walking and longeing, the swelling continued to go down. From two 15-minute longeing sessions per day, things progressed so beautifully that in less than 2 weeks from the bite, I was riding Seeker in the arena and on the trail.

Seeker has a great disposition, had thorough training, and is a very good healer. Because of that, she was able to tolerate some painful procedures, and I was able to do my very best work and help her recover quickly.

PART THREE
The Work

Your horse has no fear, and you've developed a partnership. Now it's time for the work. For each lesson, using the basics he's already learned, ask your horse to perform the new maneuver. Then reward him each time he tries to work toward the end goal. You'll always be honing because horses are never really "finished." A horse is always a work in progress.

In addition to learning the skills, a horse must get used to physical exertion, sweating, wet saddle blankets, flies, and being tied at a hitch rail for long periods. A horse needs to feel comfortable with all of the routines related to training and use.

You have a communication system; now you will refine it. In the skills phase of a horse's training, he'll learn everything he needs in order to become a great reiner, jumper, trail mount, or dressage horse.

Once a horse knows his ABCs, he's ready to form an unlimited variety of words and sentences. The basics allow you to improvise and specialize so you can develop your horse's potential and reach your goals.

Forward into Contact

Most horses have ample get-up-and-go, so we are often more concerned with stopping them. But all it takes is knowing one balky horse to see what a gift forward movement is — it is the energy we use to shape everything else.

WHEN I TOOK DRESSAGE LESSONS for several years, I was fortunate to ride under some excellent classic instructors. The voice of one, in particular, stands out in my memory because he so often gave this simple but extremely effective advice: "Go forward!" And its variation: "More forward." And that, in a nutshell, fixed many types of problems ranging from a horse losing his rhythm when learning the half pass to a horse getting too bunched up when being introduced to collection.

"Forward" is now an automatic response and a permanent part of my troubleshooting MO. I quickly learned by riding and by watching others ride that we can create some pretty complicated problems when we forget the basic tenet of riding: forward motion.

The Benefits of Forward Motion

Without forward motion you cannot have a fluid canter/lope depart, a turn on the hindquarters, or a side pass. When things get sticky, it is usually because the forward motion, the momentum, the rhythm of the footfall pattern of the gait, has been lost.

When I am working on a more collected maneuver, such as a collected canter on a 10-meter circle, I may feel my horse start to question whether he should or can do a particular maneuver. At that moment, I think "more forward" and temporarily shift to a different exercise, such as back to the regular canter or to the extended canter, to reestablish the forward movement before returning to the collected work.

The same would be true if I were riding Western. If a horse starts to bunch up while loping small circles, I might line him out down the long side of the area to open him up a bit. This not only relaxes the horse physically, allowing him to stretch out, but it takes mental pressure off both of us as well.

It is natural to think of forward motion in relation to riding, but it begins much earlier — in fact, during those first leading lessons with the young foal. We want our horses to step out briskly and with energy and impulsion. It helps if we project that we are going somewhere with our mental plan and body image. Horses pick up on our goals.

Likewise, if we have a hesitant mental picture and are looking back at the foal when leading or down at our hands when riding, instead of where we are going, the horse will pick up on that too.

From the first lessons, therefore, think forward, project forward, and teach your horse to go forward, whether it is in-hand, on the longe line, or under saddle. And don't forget to teach your horse to be led with a bridle as well as a halter.

You need to be able to move your horse forward in a variety of ways. One is drawing him forward by halter pressure (above) and another is sending him by driving, in this case with a hand signal (below).

When You Want Forward Motion

ABOVE: Forward motion begins with active in-hand work with a foal. Seeker is really stepping deep underneath herself and walking out.

LEFT: Energetic forward motion is key to ground training. Seeker is forward and balanced during her longe session.

RIGHT: Forward motion is essential when riding, as demonstrated by this dressage horse and rider.

Here are some examples where forward motion is important.

IN HAND
- Taking a horse from point A to point B
- Leading into a trailer
- Leading over obstacles
- Leading a horse at a trot in an emergency

ON THE LONGE
- All the forward gaits

WHEN RIDDEN
- Riding from point A to point B
- All upward transitions (see a complete list in chapter 15)
- Backing (see a discussion of backing in chapter 8)
- Racing, jumping, gaming
- Riding that requires extended gaits
- Crossing water, bridges, and other obstacles

Control and Contact

When we talk about forward motion, we have to discuss control, because there is nothing worse than being on a runaway. Along with the discussion of forward motion, therefore, comes the concept of contact.

A horse must learn to accept regulation (control) of his forward movement by "listening" to your seat and legs and the action of your hands through the reins to the bit. When your seat becomes still and your hands close on the reins and you no longer apply leg aids for forward movement, your horse should decrease or cease forward movement. The simplest form of this lesson is the walk-to-halt transition but includes all downward transitions as well as **shortenings** (collection) within the gaits.

Gradually, your horse must learn to accept the connection made from your seat and leg aids to the rein aids as you ask him to perform a variety of transitions and maneuvers. This is the process of a horse accepting contact, of getting him "on the bit." Contact is what binds the horse and rider together. A horse who understands and respects contact with his rider is more ready at a moment's notice to change direction or speed and is less likely to come unglued, run away, spook, or bolt.

Depending on your style of riding, your horse might need to know how to be ridden with one or two hands. All horses should accept being ridden with two hands on a snaffle bit. If you ride western, play polo, rope, or do ranch work, your horse will need to learn to neck rein so you can ride with one hand, freeing up the other to work or play. All horses should be accustomed to be ridden with one hand so that tasks common to all riders can be accomplished safely, such as riding through a gate or adjusting a hat or jacket while moving.

LEFT: Once a horse thinks forward, you merely have to guide him. Zinger steps decisively and deeply and walks straight across the platform with very little guidance from me.

RIGHT: Forward motion is key to prompt trailer loading. Zinger acts as though she is going somewhere as she loads energetically.

Contact is what binds the horse and rider together.

Forward into Contact

Test A. Lead your horse down the long side of the arena, walking as fast as you can. Then remove his halter and lead him halterless. He should perform with equal forward movement either way.

Test B. Longe your horse at an ordinary trot or jog and then ask him to reach out and extend the gait. You should see immediate propulsion from the hindquarters and a surge forward.

Test C. Lope or canter your horse from a walk. He should do so as soon as the aids are applied, and with energy. You should feel like your horse is climbing and reaching up to the lope.

10

Bending and Flexing

*We often ask horses to bend when they want to go straight and
to go straight when they want to turn.*

ORSES BEND LATERALLY AND FLEX vertically. **Bending** is a sideways arcing and **flexing** is an upward arching. The degree to which a horse can bend and flex depends on his conformation, condition, and state of relaxation. A heavily muscled horse might not be as flexible as a more willowy horse. An overweight horse often has more difficulty bending than a fit horse. A tense horse tends to be more rigid than a relaxed horse.

Lateral bending is most easily seen in the curve a horse's neck makes as he turns to the left or right. A horse's body also bends from withers to tail (the balance of his spine), but not nearly as dramatically as does his neck.

Vertical Flexion

Vertical flexing (also known as **collection**) occurs in part when a horse flexes at the poll and elevates and rounds his neck and spine. This is most evident by an elevated poll and an arch in the neck. But vertical bending also occurs in two more places. When a collected horse flexes his abdominal muscles, this raises his back. Also, when a horse steps well underneath his body with his hind legs, he bends the lumbosacral joint and the hind legs. So the composite vertical flexion picture, or collection, is an elevated forehand, arched and reaching neck, raised back, flexed loin, dropped croup, and flexed hind legs.

While lateral bending is something we begin teaching our horses from day one and continue throughout their lives, vertical flexion is more of an acquired form that a horse gains over time from physical development and balanced work. Some horses, because of their conformation, are born moving balanced and with a natural form of vertical flexion and collection, so it just needs to be further developed. Other horses have a flat topline and are heavy on the forehand; they require a lot of upward and downward transitions to help them develop vertical flexion and front to rear balance.

BELOW: Vertical flexion exhibited by a Western Pleasure horse: a rounded topline, hindquarter engagement, an elevated poll, and flexed abdominals.

LACK OF FLEXION AND COLLECTION

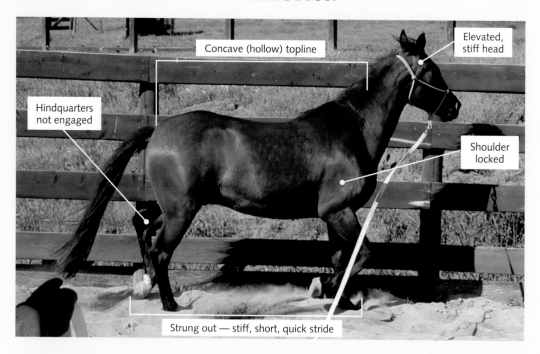

Concave (hollow) topline

Elevated, stiff head

Hindquarters not engaged

Shoulder locked

Strung out — stiff, short, quick stride

BEGINNING OF FLEXION AND COLLECTION

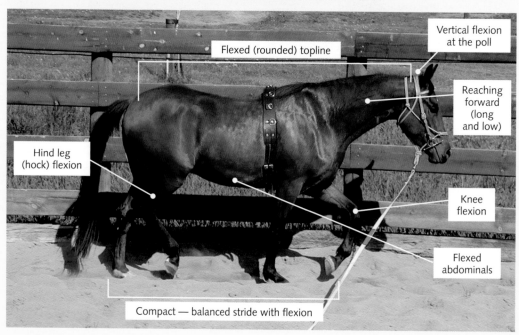

Flexed (rounded) topline

Vertical flexion at the poll

Hind leg (hock) flexion

Reaching forward (long and low)

Knee flexion

Flexed abdominals

Compact — balanced stride with flexion

Lateral Bending

Similarly, some horses are born more laterally supple because of their conformation and/or temperament. They bend very easily: some, in fact, too easily. Super supple, relaxed horses can be like rubber bands. When you ride such a horse, it is as though trying to push a rope up the road — the horse wiggles and waggles all over the place — and you wish for a little *more* resistance. Other horses are very stiff and tough and require proper conditioning and suppling exercises to help them attain a more flexible range of motion.

Most horses, however, are somewhere in between the two extremes. Throughout their training, we vacillate between exercises, asking them to bend a little more for this and to hold themselves a little straighter for that. We often ask them to bend when they want to go straight and to go straight when they want to turn. Over time, we develop a horse who we ride between the aids: a horse who continues going forward until we ask him to turn, and is ready to turn at a moment's notice.

A horse needs to bend laterally so that we can turn him, guide him, and get him where we need to go, whether we are leading or riding. We need to be able to change a horse's direction and face him in a new direction. A change of bend or change of rein is the simplest use of, and purpose for, lateral bending. Some of the maneuvers that include a change of bend are the change of rein, half turn, figure 8, and serpentine.

Beginning Bending

Successful bending requires that the horse has no fear of restriction, is willing, and has learned how to yield.

We start when leading the foal. Even though a foal has a short and sometimes stiff, upright neck, he still can bend, especially when not holding his neck rigidly up in fear. It is a common reaction for a foal to raise his head and neck when he feels pressure from a halter, but in that configuration, it is almost impossible for him to bend his neck and turn. So it is important to do some preliminary stationary exercises with halter pressure before you ask a foal to turn while striding out at the walk.

TOP: Horses are born flexible. Pete, even with his short foal neck, can turn all the way around to scratch an itch on his hindquarters.

BOTTOM: Arcing turns should be started early in a foal's training. Dickens begins his bending lessons.

There is a difference between bending and turning. **Bending** is a change in body position, usually the neck. **Turning** is change in movement involving the whole body.

A horse can bend his neck laterally at a standstill. In fact, lateral bending exercises are some of our first in-hand lessons with a young horse to establish yielding.

A horse can laterally bend his neck and spine somewhat and still go straight forward, such as when we ask for a shoulder-in. A negative example is **rubber-necking:** when we ask a horse to bend and turn and he evades our aids and bends just his head and neck but keeps his body going straight ahead.

Lateral bending is flexing the body one way or the other.

Turning, on the other hand, involves a weight shift and a change of direction. When a horse shifts his weight from one side of his body to the other and starts to leave the track he was on, he is turning. Most often, bending is an inherent component of turning. The most classic example of turning is a horse working evenly on a circle, bent from head to tail in a perfect arc of the circle with his feet working on the circle line.

Sometimes, however, we turn a horse without asking him to bend laterally very much. When we ride a horse in a spin or a canter

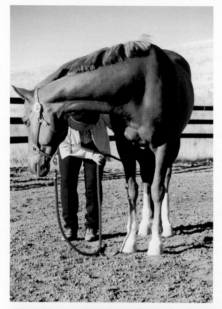

LEFT: Bending is the curving of the neck. Dickens bends around to see what I am doing at his hindquarters.

BELOW: Turning is the curving of the neck and body combined with a change of direction. Here Seeker has just bent in response to my request with my left hand and will next step her right front over her left front and turn to the left.

pirouette, we keep his body relatively straight and vertically collected as we turn. We ask him to bend only slightly at the throatlatch in the direction of the turn. The reason is that if we add too much lateral bend to such maneuvers, the horse's hindquarters will lose their position under the horse's body and come off the track.

Guidance of a horse's form and direction includes various combinations of bending and turning with forward motion.

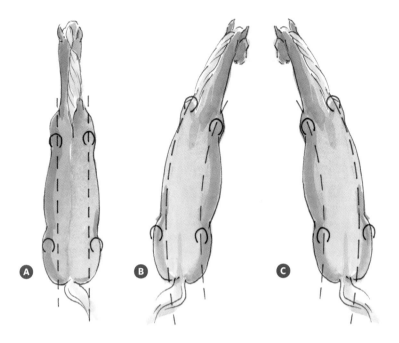

Footfall patterns of :
Ⓐ horse traveling straight
Ⓑ horse traveling to the right
Ⓒ horse traveling to the left

The same goes for an older horse. The more natural and relaxed the neck is held, the better the bending will be. In fact, if a riding horse either raises and locks his head and neck or plunges his head toward the ground, it will be physiologically difficult to turn him. There is a sweet spot for each horse — a place where the horse holds his neck and head at a natural, moderate level — where bending, and therefore steering, is easiest.

"Handedness"

We see our horse reach around with his teeth to his side or even his rump to swat a fly or scratch an itch, so we know he can bend. But you'll notice that most horses will tend to turn one way more than the other. Especially with horses in stalls, it becomes apparent that they have a habitual way of turning. This has led to the discussion as to whether some horses are right or left "handed."

Most horses do turn one way more easily and fluidly than the other, stay more balanced and upright turning one way, and will track truer on a circle in one direction than the other direction. Whether this is something inherent or is influenced by how a foal was positioned in the womb, how a horse stands to graze, is due to handling predominantly from the near side,

Bending and Leads

A horse should know how to lope or canter on both the right and left leads. The more ambidextrous and supple your horse is, the easier it will be for him to take the correct lead in either direction and lope with the proper amount of bend.

TOP: A Western horse bent evenly from head to tail at the walk. Note the deep step underneath with the right hind and the lower head carriage of a Western horse.

BOTTOM: An English horse bent evenly from head to tail at the canter. Note the elevated head and uniform bend.

or a combination of these, is hard to say, but most horses bend more easily one way than the other.

And whether going to the right or to the left, most horses travel a bit crooked in both directions because it requires less energy to let the hindquarters fall off the track one way or the other, than to keep the hindquarters up under the body.

The majority of horses, when tracking to the left in an arena or round pen, will tend to overbend into the circle — that is, a horse will carry his head low and to the inside, falling in on his left shoulder, letting more weight settle on his left shoulder and foot. The hindquarters either stay on the track or swing off the track to the outside, with the left hind leg being fairly deep under the horse's body and bearing the most weight of the hind legs. The effect is of a horse moving diagonally along, somewhat crablike.

When the same horse goes to the right, he is stiff in a different way, often holding his head and neck up and out to the left and letting his hindquarters drift off the track to the inside. This, too, resembles a diagonal crablike movement.

To change this natural tendency and get a horse to move balanced in either direction, I've found it is best to start working while tracking right. Establish contact with the outside (left) rein. While making sure the horse is going forward with good impulsion, gradually add bend to the inside (with the right rein and right leg) and get the horse to move his hindquarters back onto the track (with your right leg or a tap with a long riding whip).

Once you get something good going to the right, you'll already have a better chance of the horse going better to the left. When going to the left, counterflex the horse by establishing a fairly strong contact on the outside rein (the right rein). This, by itself, will naturally tend to straighten the horse's body by bringing his hindquarters back on the track. Eventually, you will only add a tiny bit of inside rein because a horse's tendency is to overbend in this direction anyway.

If your horse's tendency is the opposite of this, reverse the instructions.

Lateral Bending

Test A. With your horse standing straight and you standing on the ground back near the saddle area on the near side, take the rein or lead rope and bring it back toward you. Your horse should bend his neck around toward you with very little pressure and keep his head vertical and his poll at or slightly above the level of his withers. No twisting of the head, no diving, no head tossing, no moving of the feet. Repeat on the other side.

Test B. Repeat Test A while mounted.

Vertical Flexion

Test A. With your horse standing straight and you standing on the near side by his head, facing the opposite way he is facing, ask him to back up. He should first flex in the throatlatch and poll, keep his head at a natural level (poll at or slightly above the withers) and begin backing straight in a fluid motion. No head raising, nose flipping, or locking up of the movement.

Test B. Repeat Test A while mounted.

Steady and Straight

★

Be consistent. Everything is training.

★

IN HORSE PARLANCE, STEADINESS REFERS to an even temperament, a rhythmic movement, and a consistent frame. Levelheadedness was discussed in various ways in earlier chapters. A consistent frame is dependent on balance so will be discussed in chapter 13. When I use the term **steadiness** in this chapter, I'm referring to an even rhythm of movement. **Straightness** refers to the body following the head and neck, and the hind feet following the line of travel of the front feet. A horse's natural tendency is to move quickly or unevenly when he loses his balance, and he travels somewhat crooked to conserve energy, since it takes more effort to move in balance and with a steady rhythm.

Knowing that and realizing that you want to improve your horse's way of going, it is tempting to think we can simply make our horses "straighten up and fly right." Yes, we can to some degree, but it is a developmental process that takes time. Don't be in a hurry. With shaping, the slower you go, the faster you'll get there.

Basic Rhythm

The trot is an important gait for early training because most horses perform it with the best balance and cadence of all of their gaits. So it's a great starting point for both of you to work on developing a steady rhythm.

Rhythm is the tempo within each gait. A gait is steady and pure when it conforms to a precise footfall pattern that has a regular metronome-like rhythm to it. Rhythm refers not

Shaping

Shaping means using a progression to develop a behavior or a movement. Each time a horse shows an improvement toward the desired goal, he is rewarded.

The more balanced a horse is, the more rhythmic his gaits. This Morgan horse canters like a metronome.

only to the true two-beat sequence of the trot, the three-beat sequence of the canter, and the four-beat sequence of the walk, but also to the precise landing of pairs of legs within those time signatures. For example, in the trot and canter, diagonal pairs of legs are supposed to land together. If they do not, if the front lands first, then the rhythm is impure and the horse's balance has shifted to the forehand.

The regularity of a horse's steps is an essential component of riding. Each gait is like a simple musical piece. Every horse plays that piece in his own particular manner, at a particular tempo, and with his own personal expression. Many horses have one or two gaits that do not have an even, precise rhythm.

To correct an impure rhythm, first allow the horse to show you his natural rhythm, even if it is too fast or too slow. Get in tune with that and try not to hinder the horse at first by trying to slow him down, working on a head set, or worrying about crookedness. If you try to work on those things before you establish a steady rhythm, you will end up with more complicated problems. First things first.

If you have a natural sense of rhythm yourself, it will be very helpful in recognizing, establishing, and influencing your horse's rhythm. If not, you can train yourself to develop rhythm by counting aloud or picturing a metronome in your mind. Music can help you to establish or maintain your sense of inner rhythm and, if appropriate for a particular gait, can be useful in developing a horse's rhythm as well. Riding bells (bells attached to your saddle) are also helpful in hearing when things get out of sync (and as an added bonus, when out trail riding, they warn bears of your approach!).

When your horse moves in a regular rhythm, this gives you a very predictable set of movements on which to base your riding and the delivery of your aids. When you can depend on a horse to move his legs in a precise and very even two-beat, three-beat, or four-beat sequence, you will find it much easier to stay with the movement of the horse, find your seat, stabilize your hands, and apply aids. Rhythm is something that is honed in both the horse and rider through practice.

The two gaits that are most often used to develop a sense of rhythm in the horse and rider are the walk and the trot.

Riding bells can be a good aid to help you hear your horse's rhythm.

Goals for the Walk

When properly executed, the walk is a flat-footed, four-beat gait. When performed correctly, there is a very even rhythm between the feet as they land and take off in the following order: left hind, left front, right hind, right front, left hind, and so on. This will give you a slightly side-to-side motion as well as a rear-to-front motion in the saddle. A horse who really reaches but stays in a pure walk gives what is called "a rein-swinging walk".

If a horse is in a hurry, though, he may have trotting on his mind and may not settle down into an even flat-footed walk. He may seem to be walking on the tiptoes of his hind hooves, which may make him **jig** (a cross between a walk and a trot) or **pace** (where the distinct four-beat pattern has been rushed and he is moving his two right legs together and his two left legs together).

Music can help you to establish or maintain your sense of inner rhythm and, if appropriate for a particular gait, can be useful in developing a horse's rhythm as well.

Goals for the Trot

The trot is usually a very regular gait with two distinct beats. It is more stable and precise than the walk or canter. Since the trot is the most balanced gait of most horses, it is most likely the gait that can help your horse develop his rhythm and balance. At the trot, the horse's legs move in diagonal pairs in a clockwork fashion, clicking off an even one–two rhythm.

It takes strength, suppleness, and coordination for a horse to track straight. After dressage training, Zinger was able to carry herself straight even on icy footing.

The right front and left hind legs rise and fall together and the left front and right hind legs work together. You can hear, feel, and see this rhythm.

Whatever trot a horse is performing, it must be pure with the diagonal pairs of legs moving in perfect unison. If the diagonal pairs become separated and no longer land at the same time, the horse is performing a four-beat gait instead of a two-beat gait. If trotting slow and on the forehand, the front feet often will land before the hinds, causing the horse to be heavy on the forehand. The horse will appear to be jogging in front and walking behind.

Straightness

In the discussion of bending in the previous chapter, the concept of straightness came up when dealing with overbending or stiffness. When a horse overbends to the inside or is stiff to the outside, his body is crooked and his hind hooves do not follow in the path of his front hooves.

Along with proper bending work, your horse should also learn how to carry himself relatively straight. But if you try to make him absolutely straight before he has learned to move forward with energy and a regular rhythm, you might end up with a horse who has lost his impulsion and is confused. Before working on straightness, he needs to be moving sufficiently forward and be on the aids. The more forward a horse moves, the more straight he will move.

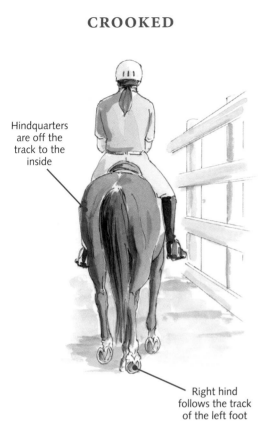

CROOKED

Hindquarters
are off the
track to the
inside

Right hind
follows the track
of the left foot

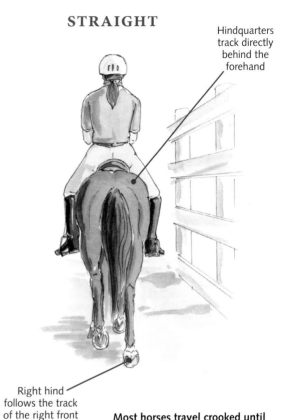

STRAIGHT

Hindquarters
track directly
behind the
forehand

Right hind
follows the track
of the right front

Most horses travel crooked until conditioned, developed, and trained. Before conditioning (left), this horse's right hind follows the track of his left front. Through the proper exercises, the horse is brought to ordinary straightness (right), where the hinds follow in the tracks of their corresponding fronts.

How Straight Is Straight?

Ordinary straightness. Every horse should learn to travel with ordinary straightness. This is where the spine is straight and the hindquarters are in a direct line behind the forehand and both of the hind legs are reaching forward under the horse's body. Because the hindquarters are typically wider than the forehand, ordinary straightness usually results in the tracks of the hinds being parallel to and slightly outside the tracks of the fronts.

Relative straightness. Advanced horses being trained to perform collected work will need to learn relative straightness. Generally, this is defined as the inside hind following in the exact track of the inside front. This is the goal because the inside hind is the major weight-bearing limb in collected movements that involve bending or turning, so it is desired to have that leg directly under the horse. The slight amount that the outside hind is off the track of the outside front is inconsequential.

★

Straightness refers to work on circles as well as on straight lines.

★

Steadiness and Straightness

Test A. With your horse on the aids, have a friend make a video of your riding.

Then ride with your horse on autopilot (self carriage).

When you play the video, use a metronome or count aloud to evaluate the steadiness of your horse's rhythm with and without aids.

Test B. Ask a friend with a photo or video camera to stand directly in front of and behind you and your horse as you perform various maneuvers. The camera should be pointed at the midpoint of the horse's chest or the midpoint of his tail head. Review the images to determine whether your horse is traveling with ordinary straightness.

Test C. With your horse haltered and on a guide line, with you holding the rope 15 feet from the halter, send your horse over an elevated platform. The horse should mount the platform dead center on the end and proceed down the middle and off the center of the other end.

Lateral Work

★

Do simple exercises well rather than advanced maneuvers poorly.

★

EXERCISES FROM THE GROUND OR saddle with some element of sideways movement are lateral movements. When a horse moves away from pressure and steps sideways, whether for a turn on the hindquarters in hand or moving over at the hitch rail, he is yielding laterally.

In early training, it is a black-and-white situation. I put my fingers on your ribs, you move over. That's the basic concept of yielding to pressure. In order for us to adjust our horse's position in the cross ties or in a trailer, he must know how to take one step at a time while performing a turn on the forehand, a turn on the hindquarters, and a sidepass.

Applications for Riding

Riding, however, introduces many more variations of lateral movement, each with its own aids and application. For example, lateral exercises vary in the ratio of forward movement to sideways movement. A sidepass has very little forward motion; it is an almost purely sideways movement. On the other hand, a "shoulder in" is mostly forward. And a half pass and two track are somewhere in between.

In lateral work, there should never be any backward movement. In general, whenever a horse performs a lateral movement, his legs should cross over and in front of each other in the direction of movement. For example, in a sidepass to the right, the left front crosses over and in front of the right front. The same is true of the hind legs.

When teaching the turn on the hindquarters, your goals are that the horse's inside hind leg (in this photo, the right hind) is a little bit farther ahead of the outside hind and is bearing the majority of the horse's weight. The horse's body is relatively straight. Overbending to the right would pop the left hind off to the left. The left front should cross over and in front of the right front. The left hind will walk a small circle around the right hind.

Lateral Essentials

It is not the object of this chapter to differentiate between all the lateral maneuvers, but rather to show their wide variations and add a few key lateral exercises to the list. Every horse needs to learn a few elementary lateral movements under saddle because they will be the basis for many practical uses and other more complex maneuvers. For example, the sidepass, turn on the forehand, and turn on the hindquarters all come in handy when opening a gate from horseback.

A leg yield at the trot. Zipper is moving to the left. He is slightly bent to the right allowing him to step deeply to the left with the diagonal pair of the right hind and left front. The left hind and right front are in flight ready to move left and cross over in front of the pair on the ground.

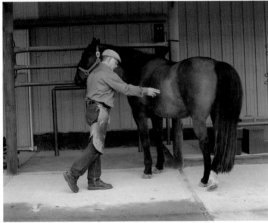

TOP: Practice the sidepass in hand before you ask for it while mounted. Seeker's right hind and left front are landing after crossing over and in front of the left hind and right front.

BOTTOM: Richard positions Seeker for farrier work by moving her hindquarters.

Lateral Maneuvers

Here is a list of many lateral maneuvers with a check mark by the exercises that every horse should know.

In hand
✔ Turn on the forehand
✔ Turn on the hindquarters
✔ Sidepass

Tied at a hitch rail
✔ Turn on the forehand

Tied in cross ties
✔ Turn on the forehand
✔ Turn on the hindquarters
✔ Sidepass

Riding
✔ Turn on the forehand
  Pivot
✔ Walk around turn on the hindquarters (haunches)
  Turnaround/spin
  Rollback
✔ Leg yield
  Two track
  In position
  Shoulder fore
  Shoulder in
  Half pass
✔ Sidepass
  Full pass
  Haunches in
  Haunches out
  Canter pirouette

Lateral Work

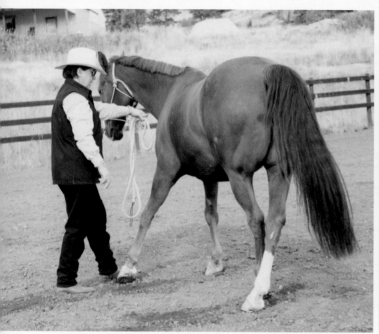

Test A. Sidepass your horse in hand. With his body relatively straight and nose tipped in slightly toward you, ask your horse to step sideways.

Test B. Sidepass your horse over a pole or railroad tie. First step him over the midpoint of the pole and sidepass off one end. Then (as shown in the photo) ride up to the end of the pole and sidepass onto and across the entire length of the pole. The horse's body should remain straight and centered on the pole with no clunks, touches, or pole rolling.

Test C. Ride up to a gate and open it from horseback. Your horse should be easy to maneuver into position so you can unlatch the gate, hold onto the gate as you open it, ride through, and close it. If you have to let go of the gate, your horse needs more work.

In this case, after the sidepass left to open the gate I'll ask Sassy to turn on the forehand (hindquarters moving right), then back up a few steps, and then sidepass left to close the gate.

Balance

★

Whether a horse is born a ballerina or a klutz, timing and balance can both be improved through exercise.

WHEN EVALUATING A HORSE'S BALANCE, we look at his static conformation and his equilibrium in motion. Balance of form refers to how weight is distributed among the body parts — forehand to hindquarters and left to right. Balanced movement is characterized by harmonious steady movement. Just as with people, some horses are born with natural balance while others must develop it. Although conformation plays a large role in balance, a poster-perfect horse that strikes a fine picture at a standstill may or may not move in balance. Beauty is as beauty does.

Proprioceptive Sense

With athletes (which horses certainly are), the **proprioceptive sense** is essential for balanced movement. This is the ability to know where various body parts are in relation to one another and whether they are moving with appropriate force and speed.

Every horse has his own innate sense of where he is putting his feet. Some are acutely aware of the position or spacing of everything on the ground and can pick their way through ground rails or brush without touching. Other horses are challenged in their timing and in knowing exactly where things are underfoot. But whether a horse is born a ballerina or a klutz, timing and balance can be improved through exercise.

A horse's conformation is his static balance. Dickens shows nice balance.

Imbalance Fore and Hind

Most horses are imbalanced in at least one way. In earlier chapters, I discussed lateral imbalances that cause a horse to overbend or move stiffly. Here, I'll focus on the imbalance of weight distribution between the forehand and hindquarters.

It is said that a horse at rest carries approximately 60 percent of his weight on the forehand and 40 percent on the hindquarters, because the forehand has the added weight of the head and neck in front of the forelimbs. But we've all seen extreme downhill horses, whose withers are several inches or more below the height of the croup. At times, those horses might bear 70 percent of their weight on their forelimbs.

Goal: Balanced Movement

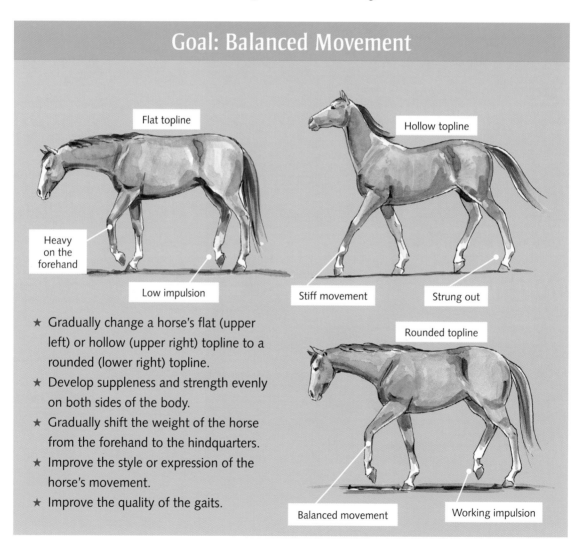

Flat topline

Heavy on the forehand

Low impulsion

Hollow topline

Stiff movement

Strung out

Rounded topline

Balanced movement

Working impulsion

★ Gradually change a horse's flat (upper left) or hollow (upper right) topline to a rounded (lower right) topline.

★ Develop suppleness and strength evenly on both sides of the body.

★ Gradually shift the weight of the horse from the forehand to the hindquarters.

★ Improve the style or expression of the horse's movement.

★ Improve the quality of the gaits.

SIGNS OF IMBALANCE

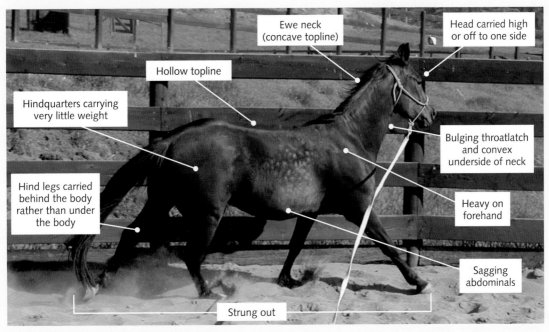

Ewe neck (concave topline)

Head carried high or off to one side

Hollow topline

Hindquarters carrying very little weight

Bulging throatlatch and convex underside of neck

Hind legs carried behind the body rather than under the body

Heavy on forehand

Sagging abdominals

Strung out

SIGNS OF BALANCE

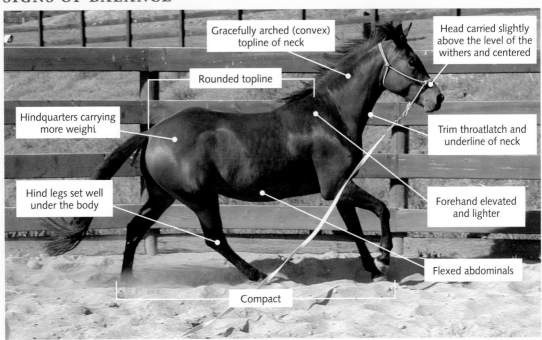

Gracefully arched (convex) topline of neck

Head carried slightly above the level of the withers and centered

Rounded topline

Hindquarters carrying more weight

Trim throatlatch and underline of neck

Hind legs set well under the body

Forehand elevated and lighter

Flexed abdominals

Compact

A horse's proprioceptive sense is his ability to know where to put his legs without being able to see where he is putting them. During a warm-up for a trail class at a show, Zinger lifts her legs high to be sure she can clear what she cannot see.

Most horses have a tendency to travel heavy on the forehand because it requires more energy to lift and shift weight rearward than it does to let the weight fall where gravity sends it, down and in front of the horse. On the other hand, there are high-headed, naturally collected horses with hind legs set well under their bodies. Those horses can shift their weight to their hindquarters and carry weight more equally between the forehand and hindquarters. All horses, however, can be brought to better balance through dedicated, systematic conditioning and training.

Benefits of Balance

Balance is essential for the horse's confidence and our safety. A horse who is discombobulated or heavy on the forehand tends to stumble more frequently. A balanced horse carries himself in a more organized package, with hind legs underneath his belly and moderate collection. Ah, there's that word: **collection**. Yes, balance has a lot to do with collection and vice versa. But true collection, as we see in upper-level dressage and reining horses, is a fairly advanced concept that is not in the immediate grasp of all horses and riders. For every horse, however, ordinary balance and at least a modicum of collection are admirable goals.

A horse who is trained and ridden in balance will develop a more beautiful physique. Once he has been developed, he will usually move and play in the pasture in a more balanced form as well. The training and conditioning changes the horse's musculature and way of going.

Balance

Test A. Lead your horse over evenly spaced ground poles at a walk and trot. Then ride him over the poles at a walk and trot. Start with the poles set 4 feet, 6 inches (1.5 m) apart and make adjustments to suit your horse. If your horse is very large, increase the distance. If your horse is very small, decrease the distance. Make note of evenness of stride and rhythm, and how many missteps, ticks, or knocks occur. Your horse should be able to maintain form and rhythm.

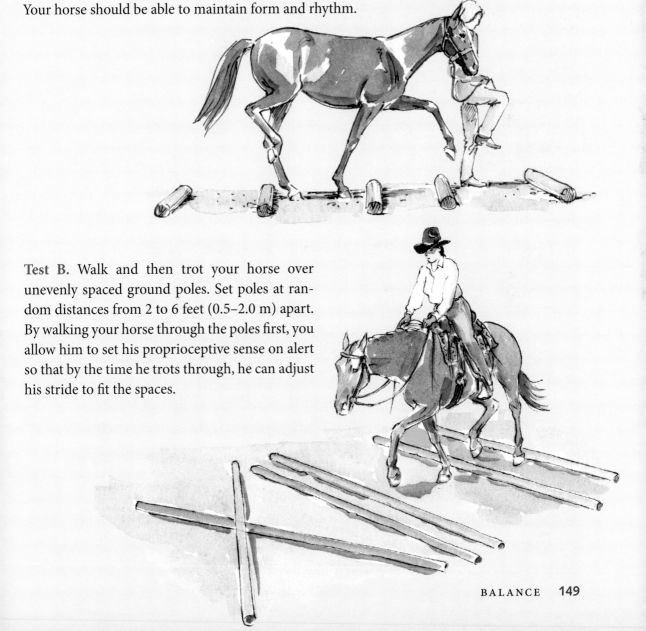

Test B. Walk and then trot your horse over unevenly spaced ground poles. Set poles at random distances from 2 to 6 feet (0.5–2.0 m) apart. By walking your horse through the poles first, you allow him to set his proprioceptive sense on alert so that by the time he trots through, he can adjust his stride to fit the spaces.

Putting It All Together

★

Two keys to a solid and versatile horse: wet saddle blankets and taking the time it takes.

★

Life is full of unpredictability, and life on horse-back is packed with its own medley of surprises. An honest, solid horse is well suited to take challenges and sudden changes in stride. To get there, though, takes time and dedication. One of the keys to a solid horse is many wet saddle blankets; in other words, there is no substitute for time under saddle and varied experience. And if you ever hope to have what is sometimes referred to as a "finished horse," or even something close to it, you need to take the time it takes — every step of the way. Mastering the basics will reward you the first time a pheasant flies up in front of you or a branch gets caught in the cinch.

Once you and your horse have mastered the concepts in this book, the rest is a piece of cake, or should I say, the frosting on the cake. Now you can use all of the skills you have developed as a team to pursue your equestrian goals.

Anytime, Anywhere

It is one thing for a horse to perform a sidepass in a quiet arena and quite another for a rider to open a gate in a hailstorm while working bawling cattle, slicker flapping, with the gate stuck in the mud. So you'll never be bored if your goal is for your horse to perform his skills anytime, anywhere. Day or night, in good weather and bad, through wind, snow, hail, rain, or mud. With commotion going on all around as well as in eerie silence. When a horse is steady and consistent in all of his skills, we call him **solid**, a horse you can count on.

When it is below zero, everything is cold: the horse's back, the saddle, the bit, the gate hinges. That's when you know that a horse is solid in his skills, as Zipper shows on this clear and brutally cold January morning.

Real-World Applications

You will have your own list of practical applications for you and your horse according to the style and use of your riding, but here is a list of real-world scenarios to show the scope of possibilities.

Equestrian Disciplines

- Pleasure Riding
- Trail Riding
- Hunting
- Jumping
- Dressage
- Reining
- Cutting
- Endurance Riding
- Driving: pleasure and endurance

- Rodeo events: barrel racing, roping, and other gymkhana events
- Drill Teams
- Racing: flat racing and harness racing
- Horse Show Exhibitions: both English and Western

Getting as much experience in hand with your young horse will pay off later when you are riding. Richard leads Sherlock back and forth calmly across a wooden bridge until he no longer reacts to the feel of the wooden planks or the hollow sound of the bridge or of the water gurgling under it.

Work

- Ranch work: rope work, carrying animals and supplies, herding, ponying, riding fence
- Farm work: pulling equipment and wagons
- Carriage work: pulling carts, wagons, and carriages in various environments
- Therapeutic riding programs: transporting riders safely while being led or ridden
- Trail work: clearing and maintaining trails, search and rescue, packing for hunting, camping, hauling supplies to remote construction sites and outposts

You never know what you are going to encounter on a trail. The more unusual things you do with your horse, the better the chance you will be able to conquer a new surprise.

Everyday Obstacles

- Going uphill
- Going downhill
- Crossing poles, rails, trees, branches
- Crossing ditches
- Crossing bridges
- Crossing water
- Going through mud
- Crossing and riding alongside roadways and traffic
- Unusual vehicles like motorcycles, bicycles, tractors, skateboards, wheelchairs, Segways
- Unusual animals: dogs, llamas, cattle
- Unusual people: children, people dressed funnily with big hats, strong perfumes

Scene Changes

- Night riding
- Snow
- Hail
- Wind
- Rain
- Fog
- Dust
- Thunder and lightning

Trail Riding

Flapping slicker (sight, sound, touch)

Wind, rain, hail, snow

Extra weight of saddle bags

Stop or slow down if caught

Brush and limbs touching head and body

Sudden appearance of wildlife

Speed control on light rein

Restriction of breast collar

Narrow passage

Irregular footing: brush, logs, rocks, holes, swales

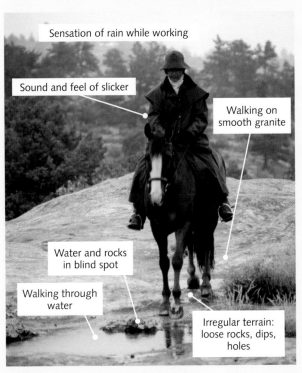

Sensation of rain while working

Sound and feel of slicker

Walking on smooth granite

Water and rocks in blind spot

Walking through water

Irregular terrain: loose rocks, dips, holes

A trail horse must be a real adventurer. All of the unpredictable surprises (bounding deer, rattlesnakes, strong wind and hail-storms, and unusual natural vistas) require a steady disposition and confidence. A trail rider and her horse need to work together as a team and not be dependent on others that might be along on the ride. The variety of footing and natural obstacles a trail horse encounters requires surefootedness and levelheadedness, as well as trust and good communication with the rider.

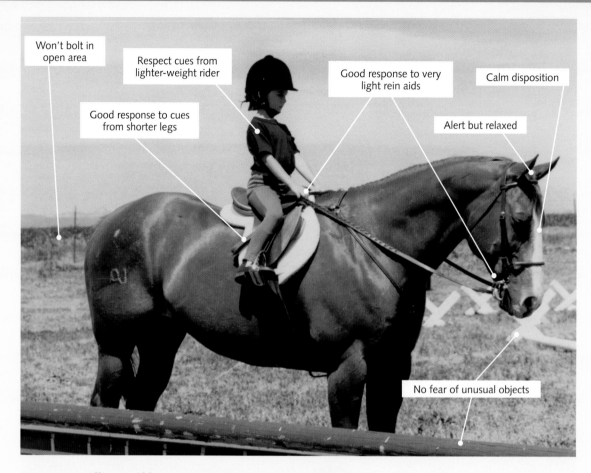

Won't bolt in open area

Respect cues from lighter-weight rider

Good response to very light rein aids

Calm disposition

Good response to cues from shorter legs

Alert but relaxed

No fear of unusual objects

Not every well-trained horse is suitable for children. A horse trained by an adult might not respond properly to a lighter-weight rider with shorter legs and less overall strength. To ensure a child's safety, a horse needs to be honest, levelheaded and calm; reliable and adaptable; responsive but not overreactive. His training needs to be very thorough so that there are no surprises.

Ponying

One-handed guiding

Sight of person above

One ear on rider, one on young horse

Dallying of lead rope to saddle horn

Calm disposition

Control on light rein

Mouth of testy young horse nearby

Pressure of breast collar if young horse pulls

A pony horse must be as well trained and bombproof as Zipper. Often the rider is paying more attention to the young horse being ponied so is not paying a lot of attention to the horse being ridden. Usually the pony horse is ridden with one hand so the rider has the other hand free to dally or control the young horse. The pony horse must be unflappable as young horses can have bursts of exuberance, laziness, stubbornness, pushiness, and more. He must go along steadily no matter what is going on alongside.

Ranch Work

Any kind of weather

Herding — keeping cattle moving

Sounds of cattle

Sight and sound of rope work

Ready to react to turning cow

Shifting gravel footing

Hard surface

Work on inclines

Potential of sudden vehicle traffic

A ranch horse has a diverse array of duties including herding, driving, sorting, roping, and carrying and doctoring cattle. To do so the ranch horse needs a steady disposition to keep the rider and cattle out of a storm (that's a mess) or a rope tangle. There are odd smells, sounds, and sights associated with branding and doctoring cattle. There are times when the horse might feel in danger such as when a big cow or bull is on the end of the rope and gets testy. That's why a ranch horse needs a certain amount of savvy or cow sense to know how to act. These ranch horses, driving a herd up a mountain road, need soundness, stamina, savvy, experience driving cattle, and no fear of passing vehicles.

Packing

Extra "dead weight" in front of saddle

Sight and sound of flapping packs

Carrying extra weight over loin

Sounds of packs moving on saddle and rubbing against brush and rocks

Sight of packs protruding from saddle

Working uphill and downhill

Variety of footing: hard, soft, wet, slippery, uneven, full of obstacles

Hard surface

Often a rider can pack enough supplies on a saddle horse for a few nights out in the wilderness, but sometimes a pack horse is needed. In either case, when you add packs to a horse's load, you're asking him to carry dead weight. While you are able to adjust your position when going uphill and downhill, the packs just sit there. Heavy packs over an unconditioned horse's loin area can make those muscles sore — that portion of the spine is not designed to carry weight. When you trot or lope, if the packs are not properly tied down they can bounce around, thumping the horse. Packs can add extra width to your overall clearance when going through trees or brush, so a pack horse needs to sense how wide his load is and compensate for it.

Goals

★

Break large lessons into smaller, achievable goals.

★

WHEN WORKING WITH HORSES, YOU will have both subjective and objective goals. **Subjective goals** are those qualities that can be discerned but not easily quantified. They include such attributes as a willing attitude, cooperation, trust, and respect. **Objective goals** can be specifically measured. They usually involve the performance of specific skills. These include achieving a certain speed, cantering on the correct lead, clearing a 4-foot fence, or standing still while a rider puts on a slicker.

Subjective Goals

Subjective goals are the foundation of horse training and, when you come right down to it, they are the more important of the two goal types. Subjective goals should never be overlooked or forsaken just to achieve an objective goal.

Subjective goals are sometimes difficult to monitor, especially for the novice horse owner or trainer. When a behavior change results in an improvement in attitude, that's great. But when subtle negative behavior changes go undetected until a very obvious change has occurred, then we have a problem.

No matter what you use your horse for, your goal should be to gain his trust and cooperation. Sassy is a "where to, what's next?" kind of horse with a great working attitude.

All too often, shifts in cooperation and trust go south. If a person is determined to make a horse back up quickly, straight, and energetically in one day and the lesson is repeated over and over, pretty soon the horse might lose his desire to go forward. Or if a competition is looming and a horse is worked hard every day for 2 weeks, a horse who used to meet you at the gate might be hanging out in the far corner of the pen when you come to catch him. That is why, as horsemen and trainers, we must be ever vigilant and never sacrifice our horse's willing attitude simply for a performance accomplishment.

Horses are individuals and each will tell you when the schedule is okay and when it is too intense. I have some horses who would be content to go round and round in an arena, working on all sorts of maneuvers *ad infinitum*. They are happy as clams doing that sort of work. Other horses, however, get bored more quickly with arena work and really brighten up when those same exercises are practiced out in a pasture, along a road, or behind some cows.

Some horses can be on a moderate-to-active work schedule 12 months a year every year, while others do better when given a break every 3 or 4 months. These individuals will be rejuvenated after just a few weeks of down time in the pasture, hanging out with other horses.

> *As horsemen and trainers, we must be ever vigilant and never sacrifice our horse's willing attitude simply for a performance accomplishment.*

Objective Goals

Objective goals are the "to-do" lists of horse training. These are the specific skills we want our horse to learn. It is usually easy to determine whether a horse has or has not achieved an objective goal — he can either perform the skill or not. For example, he can lope on the left lead from the walk or he cannot. But in many maneuvers, the matter of form or quality of performance enters in and designates the degree of a horse's achievement. If a horse lopes but takes a few trot steps in between the walk and lope, he has not reached the final goal, but he is close.

Here are some checklists to help you plan and gauge your horse's training.

Foal

NO FEAR TOUCH ALL OVER

☐ Wither scratch

☐ Tail head scratch

☐ Ears

☐ Mouth

RESTRAINT

☐ Hold by arms

☐ Catch easily

☐ Halter and unhalter smoothly

☐ Can be turned loose safely without bolting away

☐ Pick up feet

☐ Butt rope for leading

SACKING OUT

☐ Familiar with blankets, cloths, grooming tools, other barn items

IN HAND

☐ Walk

☐ Trot

☐ Stop

☐ Turn left

☐ Turn right

☐ Back

☐ Turn on the forehand

☐ Turn on the hindquarters

☐ Stand on a long line

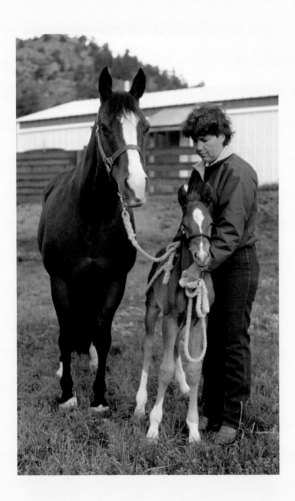

A foal's lessons should begin very early. With Sassy along his side, week-old Dickens is ready to go on a walkabout.

Foal *(continued)*

Soon a foal should be worked solo.
Throughout the suckling period,
Sherlock and I went on walks
farther and farther from his tied
dam Sassy.

VARIATIONS FOR ALL IN-HAND WORK

☐ Alongside dam

☐ Behind dam

☐ In front of dam

☐ Away from dam with dam tied

☐ From near and off sides

HANDLING BY VET

☐ Mouth handling

☐ Tail and anus area handling

☐ Leg handling

☐ Examination of all parts of the body including belly, sheath, udder

HANDLING BY FARRIER

☐ Picks up and holds up each foot for hoof care

TYING

☐ Stands quietly while tied near dam

☐ Moves over while tied

OBSTACLES

☐ Can be easily led through a gate

☐ Can be easily loaded in a trailer

☐ Can be hosed down with warm water and cold water

Weanling

☐ Review all foal exercises

TYING

☐ Stands in cross ties

OBSTACLES — LEAD OVER

☐ Ground poles (not raised)

☐ Plywood or platform bridge

☐ Concrete

LEAD NEXT TO

☐ Flag

☐ Vehicles or farm equipment

☐ Objects on fence

SACKING OUT

☐ Clipping

☐ Bathing

☐ Sacking to blankets

The weanling needs to learn many things in a short time. Trust in the handler, as Blue shows here, is the key to mastering the lessons.

Yearling

☐ Review all foal and weanling exercises

ADDITIONAL OBSTACLES

☐ Ground poles raised

☐ Tarp or plastic

☐ Lead over ditches, up and down banks

ADDITIONAL SACKING OUT

☐ Whip

☐ Ropes all over: head, legs, chest, under tail

☐ Plastic and paper bags all over

TACK

☐ Fit and carry bit and bridle

☐ Blanket

☐ Surcingle

A long yearling is ready to be introduced to work. Little Red Man starts out with some free longeing in a 65-foot round pen.

Two-Year-Old

☐ Review all foal, weanling, and yearling
    exercises

TACK

☐ Saddle

☐ Front and back cinch

☐ Breast collar

☐ Crupper

☐ Boots

GROUND WORK

☐ Longeing

☐ Ground driving

By the time a horse is two years old, he should have
been saddled and bridled and be familiar with the
sensations of that tack. Casper goes through his paces,
fully tacked up and with the reins lightly tied to the
saddle horn.

Long Two-Year-Old

Long means approximately 18 months old;
in the fall of the 2-year-old year.

FIRST RIDES

☐ Stand for mounting

☐ Move forward

☐ Stop

☐ Stand

☐ Turn both ways

☐ Stand for dismounting

When it is time to step up for that first ride, there
should be few new sensations. Dickens is wondering
what is going on up there, so I give him a loose rein
and urge him to move forward.

Three-Year-Old

☐ Review all foal, weaning, yearling, and two-year-old exercises

☐ Riding

☐ Begin Transition Mastery (see Transitions box)

☐ Begin Maneuver Mastery (see Maneuvers Box)

As the young horse becomes more solid in his training, he should be exposed to new things. Although Zipper is only green broke here, I welcomed the opportunity to ride him while there was heavy equipment working near the arena in order to broaden the scope of our experiences together.

Transitions

Transitions are the shifting of gears, up and down, between and within gaits.

BASIC

☐ Halt–Walk

☐ Walk–Halt

☐ Walk–Trot

☐ Trot–Walk

☐ Halt–Trot

☐ Trot–Halt

☐ Trot–Canter/Lope Right Lead

☐ Canter/Lope Right Lead–Trot

☐ Trot–Canter/Lope Left Lead

☐ Canter/Lope Left Lead–Trot

☐ Walk–Canter/Lope Right Lead

☐ Canter–Lope Right Lead–Walk

☐ Walk–Canter/Lope Left Lead

☐ Canter–Lope Left Lead–Walk

ADVANCED

☐ Halt–Canter/Lope Right Lead

☐ Canter/Lope Right Lead–Halt

☐ Halt–Canter/Lope Left Lead

☐ Canter/Lope Left Lead–Halt

☐ Walk–Extended Walk

☐ Extended Walk–Walk

☐ Trot–Extended Trot

☐ Extended Trot–Trot

☐ Canter/Lope Right Lead–Extended Canter/Lope Right Lead

☐ Extended Canter/Lope Right Lead–Canter/Lope Right Lead

☐ Canter/Lope Left Lead–Extended Canter/Lope Left Lead

☐ Extended Canter/Lope Left Lead–Canter/Lope Left Lead

☐ Walk–Collected Walk

☐ Collected Walk–Walk

☐ Trot–Collected Trot

☐ Collected Trot–Trot

☐ Canter/Lope Right Lead–Collected Canter/Lope Right Lead

☐ Collected Canter/Lope Right Lead–Canter/Lope Right Lead

☐ Canter/Lope Left Lead–Collected Canter/Lope Left Lead

☐ Collected Canter/Lope Left Lead–Canter/Lope Left Lead

Maneuvers

BASIC

- ☐ Half Turn Right at the Walk
- ☐ Half Turn Left at the Walk
- ☐ Full Circle Right at the Walk
- ☐ Full Circle Left at the Walk
- ☐ Figure Eight at the Walk
- ☐ Serpentine at the Walk
- ☐ Half Turn Right at the Trot
- ☐ Half Turn Left at the Trot
- ☐ Full Circle Right at the Trot
- ☐ Full Circle Left at the Trot
- ☐ Serpentine at the Trot
- ☐ Figure Eight at the Trot
- ☐ Full Circle Right at the Canter/Lope
- ☐ Full Circle Left at the Canter/Lope
- ☐ Turn on the Forehand Right
- ☐ Turn on the Forehand Left
- ☐ Turn on the Hindquarters Right
- ☐ Turn on the Hindquarters Left
- ☐ Leg Yield Left at the Walk

- ☐ Leg Yield Right at the Walk
- ☐ Leg Yield Left at the Trot
- ☐ Leg Yield Right at the Trot
- ☐ Leg Yield Left at the Canter/Lope
- ☐ Leg Yield Right at the Canter/Lope
- ☐ Sidepass/Half Pass at the Walk
- ☐ Figure Eight at the Canter/Lope with Simple Lead Change
- ☐ Serpentine at the Canter/Lope with Simple Lead Change

ADVANCED

- ☐ Half Turn Right at the Canter/Lope with Counter Canter, Change of Lead or Drop to Trot
- ☐ Half Turn Left at the Canter/Lope with Counter Canter, Change of Lead or Drop to Trot
- ☐ Figure Eight at the Canter/Lope with Flying Lead Change
- ☐ Serpentine at the Canter/Lope with Flying Lead Change

Aged Horse

☐ Review foal, weaning, yearling, two-year-old, and three-year-old exercises

RIDING

☐ Continue adding and honing transitions and maneuvers

☐ Add specialized skills or maneuvers for
   your sport or work, such as:

☐ Jumping & cross country

☐ Sliding stop & other maneuvers

☐ Carrying a calf

☐ Collected movement

☐ Dragging items with a rope

☐ Ponying

☐ Galloping

☐ Riding double

☐ Vaulting

The aged horse comes in handy for all sorts of uses
including ponying. Zinger, although a mare and not
the usual choice for a pony horse, tunes out the antics
of young gelding Sherlock, who by this time is taller
than she.

Scale of Training

When training horses there is so much overlap, so much going on at one time, that it is difficult to say which step should be discussed first. I've often said that a horse needs to learn how to "go forward, steady, straight, turn, and stop" and in a nutshell, that is true. All of the steps have equal importance, relatively speaking, and yet, for safety sake, you might put more emphasis on stop, while a dressage rider might put more emphasis on forward motion.

To help you to develop your own program, in addition to my discussions, you should consider the classic dressage scale of training. Dressage means training and dressage usually refers only to mounted training but still can provide valuable insight into how goals are interrelated in stages.

The scale is divided into three phases. You'll see that certain elements appear in a phase, but not just in that one phase. There is overlap.

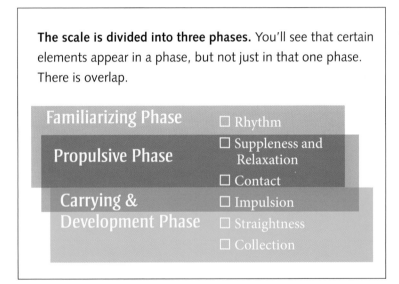

Familiarizing Phase	☐ Rhythm
Propulsive Phase	☐ Suppleness and Relaxation
	☐ Contact
Carrying & Development Phase	☐ Impulsion
	☐ Straightness
	☐ Collection

Movement Terminology

Rhythm. Traveling with a steady footfall pattern of the gait.

Suppleness. Flexibility laterally and vertically.

Relaxation. Being unconstrained mentally and physically.

Contact. Moving up to reach the pressure on the bit and the bridle in response to the rider's other aids.

Impulsion. Thrust from the hindquarters sent through the back to the neck.

Straightness. Keeping the body and legs on the track.

Collection. Engagement of the limbs and spine so that the horse is in balance with supporting hindquarters, lowered croup, and elevated forehand.

Afterword

Chewing on the things I have written about in this book made me reflect back to something I said early on. A horse is born with everything he needs. In fact, it seems that some horses are born trained. Things just come naturally and easily to them. That means the more you think like a horse, the easier it will be to stay out of the way of his success. It still amazes me when I think of all of the mistakes a novice might make and yet how the horse remains forgiving and willing to try again. If you are a mindful horse trainer, just think of the possibilities.

I wish you the very best of success working with your horse!

Index

Other Storey Titles You Might Enjoy

101 Arena Exercises for Horse & Rider, by Cherry Hill.
Classic exercises, suitable for both English and Western riders, along
with the author's own original patterns and maneuvers.
224 pages. Paper with comb binding. ISBN 978-0-88266-316-6.

Cherry Hill's Horsekeeping Almanac.
The essential month-by-month guide to establishing good routines
and following natural cycles to be the best horsekeeper you can be.
576 pages. Paper. ISBN 978-1-58017-684-2.

The Horse Behavior Problem Solver, by Jessica Jahiel.
A friendly, question-and-answer sourcebook to teach readers how to
interpret problems and develop workable solutions.
352 pages. Paper. ISBN 978-1-58017-524-1.

The Horse Training Problem Solver, by Jessica Jahiel.
Basic training theory and effective solutions and strategies in a handy
question-and-answer format — the third book in a popular series.
416 pages. Paper. ISBN 978-1-58017-686-6.

The Horse Conformation Handbook, by Heather Smith Thomas.
A detailed "tour of the horse," analyzing all aspects of conformation
and discussing how variations will affect a horse's performance.
400 pages. Paper. ISBN 978-1-58017-558-6.

Horse Handling & Grooming, by Cherry Hill.
A wealth of practical advice on mastering dozens of essential skills
necessary to maintaining equine well-being.
160 pages. Paper. ISBN 978-0-88266-956-4.

How to Think Like a Horse, by Cherry Hill.
Detailed discussions of how horses think, learn, respond to stimuli,
and interpret human behavior — in short, a light on the equine mind.
192 pages. Paper. ISBN 978-1-58017-835-8.

These and other books from Storey Publishing are available
wherever quality books are sold or by calling 1-800-441-5700.
Visit us at *www.storey.com.*